Tools and Processes

A. Yarwood

HODDER AND STOUGHTON
LONDON SYDNEY AUCKLAND TORONTO

Preface

This is the fourth in a series of books by the same author. All four books have been written as textbooks for those who are aiming to gain a certificate in one of the examinations in subjects taught in Craft, Design and Technology (CDT) departments in schools and colleges. The books are suitable for preparing pupils and students in fourth, fifth and lower sixth forms for 16+examinations in CDT subjects. The first three of the series of books – *Design and Craft* (written in collaboration with S. Dunn), *Design and Technology* (written in collaboration with A. H. Orme) and *Design and Woodwork* are all centred around a defined design process which is now being used with much success in many educational establishments. This fourth book, *Tools and Processes*, deals with the technicalities of working those woods, metals and plastics commonly found in Craft, Design and Technology departments. This present book can therefore be regarded as an extension to each of the other three in the series. Whereas the other three deal with designing, this fourth volume deals with the tools and the processes which are necessarily involved in the making (or realisation) of the designs which pupils and students have originated.

Books by A. Yarwood

Design and Craft by A. Yarwood and S. Dunn
Design and Technology by A. Yarwood and A. H. Orme
Design and Woodwork by A. Yarwood

Teach Yourself Woodwork by A. Yarwood
Teach Yourself Graphical Communication by A. Yarwood

Acknowledgements

The line drawing of a strip heater which appears on page 36 was copied from a photograph in a catalogue kindly supplied by C. R. Clarke and Co. of Ammanford, Dyfed, Wales.
The series of photographs and text on pages 46 and 47 were taken from a wall chart *Aluminium Sand Casting* published by William Allday and Co. Ltd., the makers of the 'Alcosa' range of equipment. William Allday and Company Limited not only supplied me with the photographs but also kindly allowed me permission to reproduce the text from their wall chart.
The photograph of a Denford TDS 6WL Wood Lathe which appears on page 53 was kindly supplied by Denford Machine Tools Limited.
The tools shown on the cover were kindly lent by The Hardware Centre, Sevenoaks, and Toolmail. The cover photograph was taken by Melvyn Gill.

ISBN 0 340 36472 6

First printed 1985

Typeset in 10/11 pt. Univers (Monophoto) by
Macmillan India Ltd., Bangalore 25.

Printed and bound in Great Britain for
Hodder and Stoughton Educational,
a division of Hodder and Stoughton Ltd.,
Mill Road, Dunton Green, Sevenoaks, Kent
by Page Bros (Norwich) Ltd

Contents

CRAFT DESIGN AND TECHNOLOGY

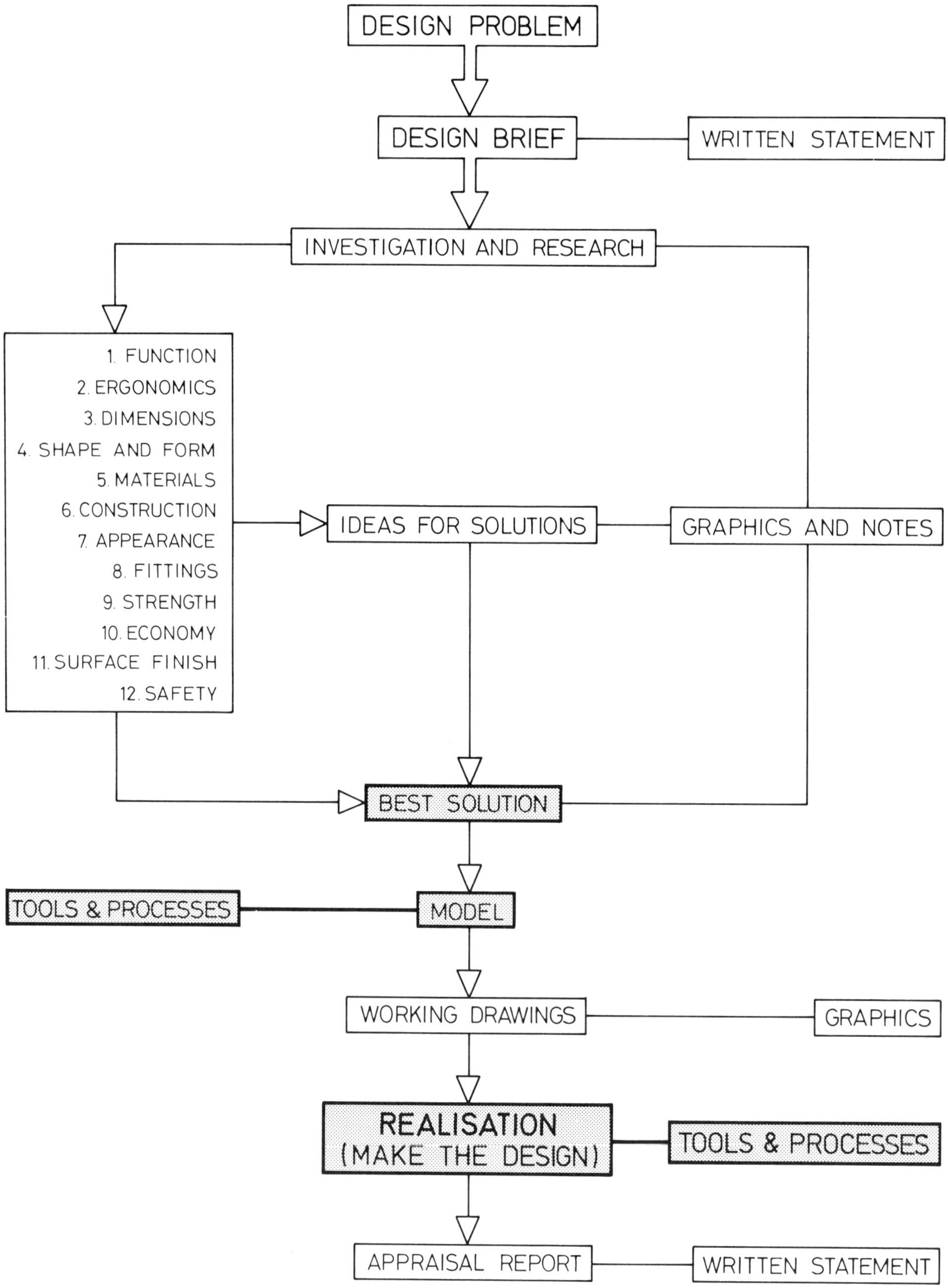

Introduction

Craft Design and Technology

This volume is one of four in a series of books for Craft Design and Technology subjects. The other three books in the series are *Design and Craft, Design and Technology* and *Design and Woodwork*. The flow-chart shown on page 4 opposite has been included in each of the books in the series. The flow-chart appears in a slightly different form in each book. It has been repeated here to show where the tools and processes outlined in this book are applied when practising Craft Design and Technology. It is in the realisation of the best solution, in the actual making of the completed design and any models associated with the design, that tools must be used in processes which shape and form, construct and finish the item which has been designed.

With the introduction of many new, modern materials, the practice of making designs wholly from just one material has given way to designs made from two or more materials. The three materials most often associated with craft and technology are metals, plastics and woods. This book describes the more common of the tools and practices involved in working in these three groups of materials.

A design process

The following is an outline description of the flow-chart on page 4 opposite. Other books in the series describe the design process in great detail.

Design problem

Designs are for solving the problems which arise from the needs, wishes and desires of human beings. When attempting the solution to a design problem, it is as well to write a statement defining what you consider to be the design problem which you wish to solve.

Design brief

After recognising the design problem state as briefly, but as clearly as possible, what it is you intend to design. This is the 'design brief'. Never attempt to solve a design problem until you have set down this written statement.

Investigation and research

Consider all the problems and difficulties which can arise under each of the twelve headings shown as 'Investigation and research'. As the investigation proceeds, write notes and draw diagrams – keep a record of the results of your investigation and research. In craft and technology design work this record will often take the form of annotated drawings, but it could equally well be in the form of notes interspersed with diagrams.

Ideas for solutions

During your investigation and research, ideas for solutions to the design brief should be drawn. Again, it is essential that as solutions come to mind they are recorded, preferably in drawing form. If not recorded as ideas occur, solutions may be forgotten and lost.

Best solution

As a number of ideas for solutions to the design brief build up, a clearer idea of what you consider to be the best solution will emerge. Make detailed drawings, with notes, of what you consider to be the best solution.

Model

Making a model of your suggested best solution will often be of value for showing whether your suggestions are suitable or will work well. If a model is to be made a selection of appropriate tools and processes will be necessary. When the model has been made, it may show that modifications are necessary in the design which you consider as your best solution.

Working drawing

When satisfied that your best solution is as good an answer to the design brief as you can provide, a working drawing from which the design can be made should be drawn. A working drawing should contain all the necessary information allowing you to make the design, without having to refer to your investigation and solution drawings and notes.

Realisation

Now the design can be made and it is here that you will need to make a good selection of appropriate tools and processes. Although this stage of designing does not appear to take much space in the flow chart on page 4, making the design will often require a greater length of time than all the other stages in the design process.

Appraisal

Always make a written report about your completed design. Do you think the design is successful? If so, why? If not, why not? Could it have been improved? How? Be honest and do not report favourably upon a design just because you have made it. An honest, clear and critical appraisal of design work will assist you in obtaining a better design for your next piece of work.

Benches and bench equipment

Benches

Benches need to be strong, sturdy and rigid. They are best made from tough hardwoods such as beech or birch. In schools, woodwork benches are usually designed to allow two pupils at each bench, working on opposite sides, facing each other. The benches are equipped with a vice on each side, one for each pupil. More advanced styles of bench have end vices fitted. A recess along the centre allows tools in current use to be placed on the bench so as not to interfere with large pieces of wood being worked. Woodwork vices are often made from malleable cast iron and steel, with wooden jaws fitted inside the metal jaws to ensure that wood held in the vice is not damaged by the metal jaws.

Woodwork benches can be adapted for light metal-work or for work with plastics by fitting a purpose-made cover on the bench top as illustrated. This is held firmly in position in the bench vices. Small *engineer's vices* (with, say, 100 mm wide jaws) fitted to this cover allow metal or plastic materials to be worked. When working with glass fibre a further protective cover of thick polythene sheet (say 1000 gauge) is necessary. Polyester resins will not adhere to the polythene.

Less space is required when working in metals. Thus four or six-place metalwork benches are common. These have tops fitted with protective angle-iron edges. Engineer's vices are the most common form of metalworking vice and those with 100 mm or 125 mm jaws are suitable for school work.

Bench equipment

Engineer's vice for cramping on woodwork bench

To enable light metalwork to be carried out at a woodwork bench, a small engineer's vice, with, say, 100 mm jaws, can be bolted to a piece of 40 mm thick wood with angle-iron along one edge. This can then be cramped to a woodwork bench when required.

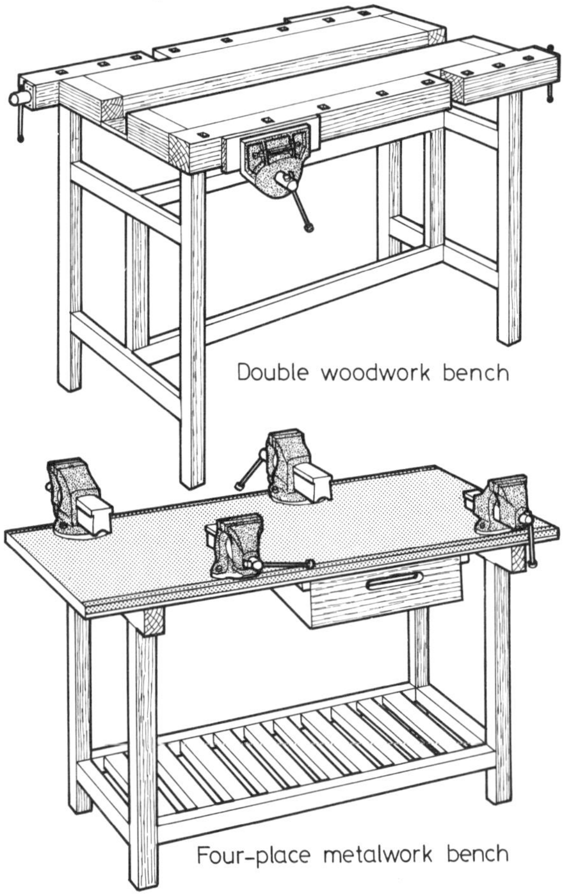

Examples of craftwork benches

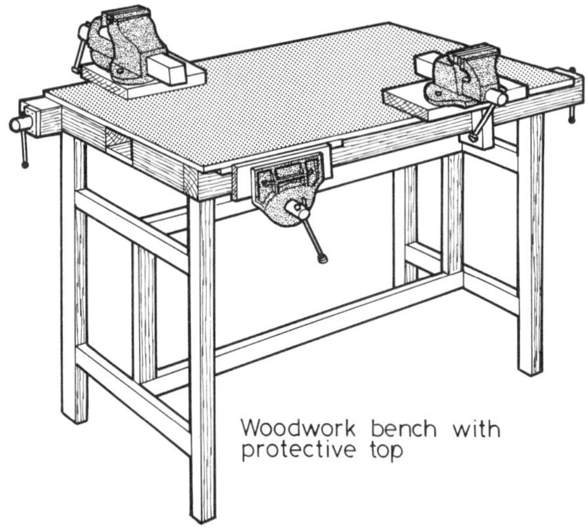

Engineer's vice for cramping on woodwork benches

The bodies of engineer's vices are made from fine cast iron. Vice screws and other working parts are made of steel and vice jaws are of hardened steel. 'Quick-grip' engineer's vices can be purchased with quick-release devices which enable the vice jaws to be opened and closed speedily.

Engineer's vices are fitted to bench tops with bolts, nuts and washers. The hardened steel jaws can be covered with fibre grips or with workshop-made vice-clamps of wood or soft metals – e.g. aluminium, lead, copper. The use of vice-clamps minimises damage to material held between the serrated steel jaws of the vice. Do not hammer on the top of the sliding parts of a vice. Such hammering can cause damage to the vice. Use an *anvil* (see page 39) or a *bench block* to hammer on.

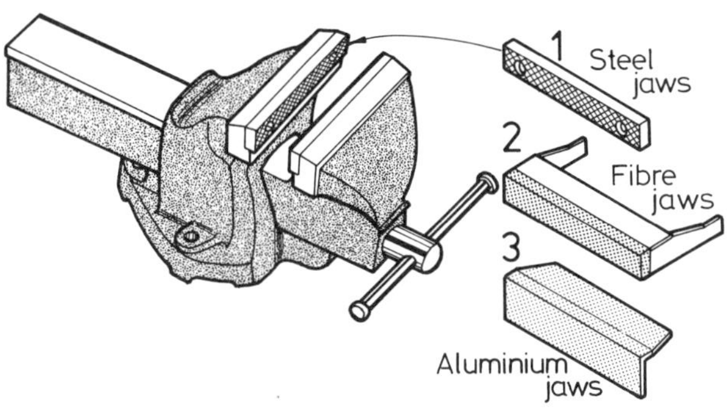

Engineer's vice and vice-clamps

Bench-stops

Woodwork benches are fitted with *bench-stops* against which wood can be held, as when planing. Some benches are made with a series of holes in the top near to the front edge of the bench, in which steel (or wooden) bench-stops will fit. Other benches are fitted with wooden stops in one position.

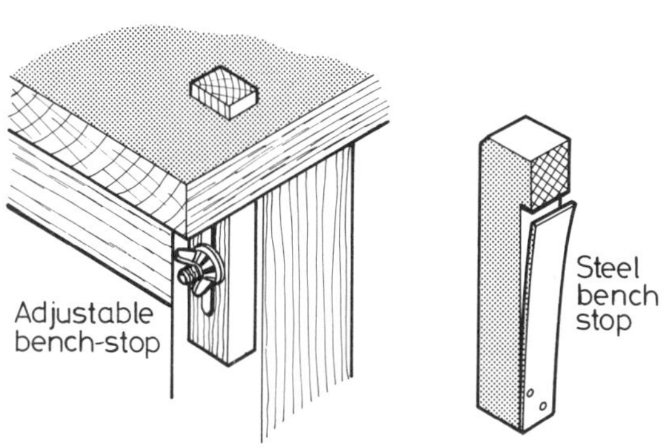

Woodwork bench-stops

Woodworking jigs

A number of *woodworking jigs* are available in which wood can be firmly held while it is worked to shape. Three of the more common types are illustrated.

Sawing board

Designed for holding wood securely with one hand while a *tenon saw* is worked with the other.

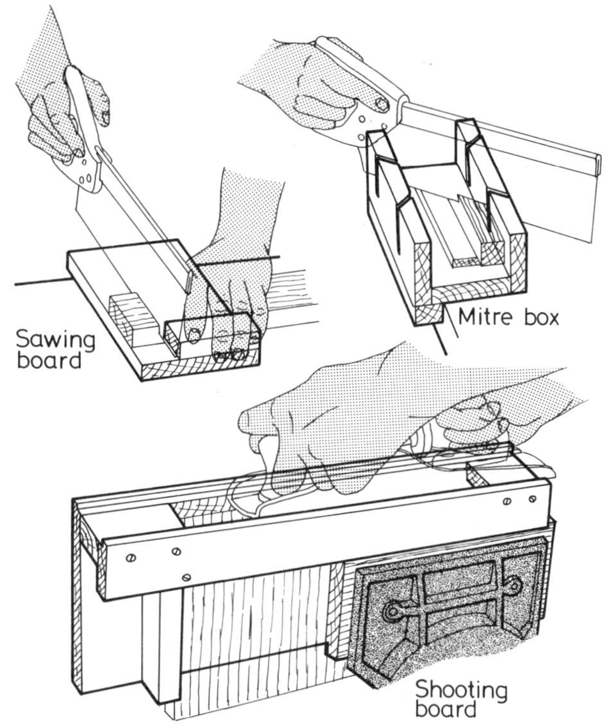

Woodworking jigs

Mitre box

This is a sawing board in which the saw is held at an angle of 45° to achieve accurate sawing of *mitre* joints.

Shooting board

Wood can be held firmly in a *shooting board* while its ends are planed accurately square.

'Jointmaster' jig

A number of proprietory jigs can be purchased for assisting in sawing square and mitred ends to wood. One such jig is the 'Jointmaster'.

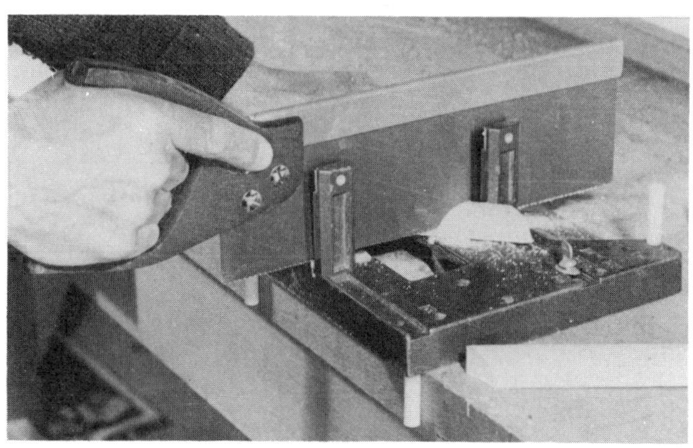

A 'Jointmaster' for sawing accurate mitres

Metalwork clamps and vices

A number of clamps and vices designed for holding small pieces of metal as they are worked are shown in drawings.

Toolmakers clamps
These are for holding pieces of metal together as, for example: they are being marked for size; when identically placed holes are to be bored in several pieces; when several pieces are to be filed to identical shapes. The two screws of the clamp allow the steel jaws to be adjusted so as to grip the materials being held firmly, while keeping the jaws parallel to ensure maximum grip along their length. Toolmakers clamps are made in a range of sizes with jaws of 50 mm length and upwards.

Hand vices
These are made in a number of lengths up to about 150 mm. A hand vice can be held in one hand while work is firmly clamped between its jaws. It enables larger pieces than can be held in pin vices to be worked in this way. One hand holds the vice whilst the second hand is free to use a hand tool. Hand vices are also of value for holding small pieces to be ground at a grinding machine, or drilled in a drilling machine.

Pin vices
Pin vices are hand vices; two types are shown. One is for holding pieces of metal of small diameter; the second for holding small pieces of any required shape. After fixing the piece being worked in the *vice chuck* or *jaws*, the vice can then be held in one or both hands to enable the small items of metal to be worked either with hand tools or at a machine.

Folding bars
These enable sheet metal to be held firmly while the sheet can be bent along the straight line formed by the top of the bars.

Stakes
A large number of differently shaped *stakes* are made for forming the shapes required when working sheet metal. The upper surfaces of stakes are polished to a fine, smooth surface to allow the sheet materials being formed on their surfaces to remain unblemished. The stakes are held in *bench sockets* bolted firmly to stake benches. Two bench sockets are shown here in photographs. Each bench socket has a square-tapered hole through it, into which the square-tapered shaft of a stake can be placed. The stake is then held firmly in the socket and a wedge-shaped taper steel key may be needed to remove it when another stake is to replace it in the socket. To do this, a rectangular slot is cut in the side of the socket or in the side of the stake shaft. The wedge of the release key is then hammered into the slot, lifting the stake from the socket hole no matter how tightly the stake has become wedged. Other stakes are shown in illustrations on page 37.

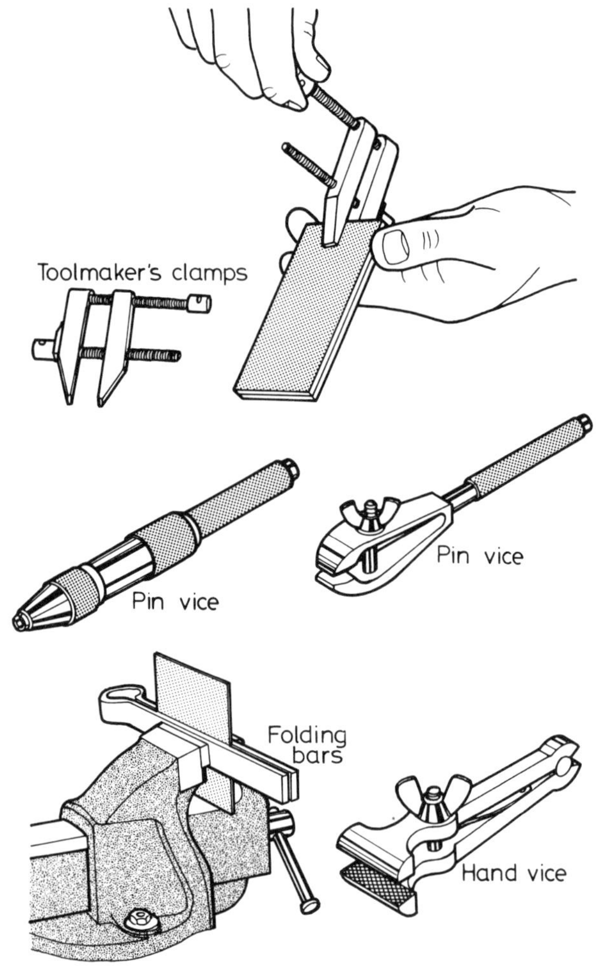

Various metalwork clamps and vices

A funnel stake and a planishing stake

Various woodwork holding devices

Apart from woodwork bench vices such as those illustrated on benches (see page 6), a very large number of different types and sizes of holding devices are made for holding wood firmly in position as it is being worked, or for holding parts of constructions together while glue sets between them. When holding wood in these devices it is important to note that pieces of scrap wood (preferably soft wood) should be placed between the wood being held and the steel jaws of the holding device. If this precaution is not observed, the surfaces of the wood being held can become severely bruised or otherwise damaged by the steel jaws.

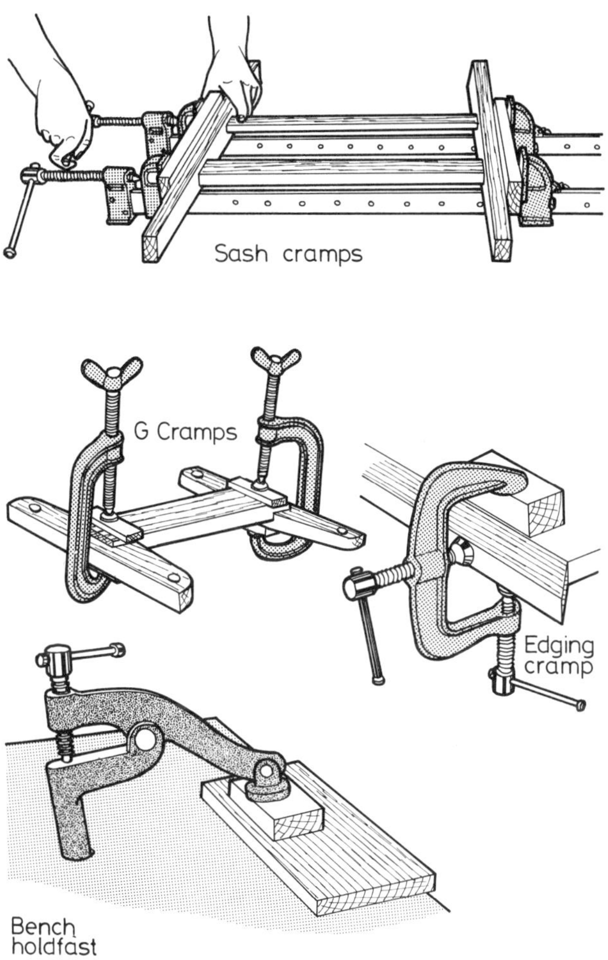

Various woodwork holding devices

Sash cramps

Also referred to as *bar cramps, sash cramps* are designed for holding together parts of wooden constructions after glue has been applied to the joints. The cramps enable the parts to be held firmly against each other until the glue sets or hardens, when the cramps can be removed. Note that, to ensure the squareness of a construction which is being cramped, the cramps should be tightened along the line of the part to which pressure is being applied.

Sash cramps are made with bars which are either rectangular or T-shaped in section and they vary in length from about 600 mm to as long as 1800 mm. Furthermore, bar extensions can be added, to allow very large constructions to be cramped.

G Cramps

G cramps have two main uses – for holding wood in position as it is worked with tools, or for holding parts together until the glue between them sets. There are a number of different types of G cramp available and they are among the most versatile of holding devices. Some, for example will handle work of sizes ranging from 50 mm up to about 300 mm. They may be made with 'Quick-grip' mechanisms (not illustrated), which enable the jaws to be opened and closed very quickly.

Edging cramp

This is one of the varieties of G cramps available. It is designed for holding *lippings* as they are glued to the edges of a board.

Bench holdfast

Pieces of wood can be held down firmly on a bench top while being worked with a *bench holdfast*. The stem of the holdfast is held in a purpose-bored hole cut in the bench top. Some holdfasts are designed to be held within a hole in a steel collar screwed into the bench top.

Corner cramp

Corner cramps are designed for holding corner mitres of frames while the glue between the mitres sets. They are usually supplied in sets of four.

Web clamp

This is a proprietory cramping device made by Stanley Tools. The nylon webbing is tightened by a ratchet device in a steel body worked by a screwdriver.

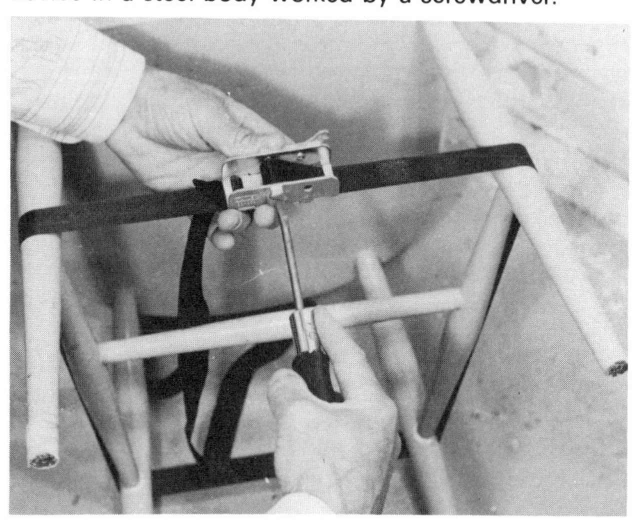

A web clamp

Marking tools and marking-out

Marking tools for wood and metal can equally well be used when marking-out plastic sheet materials such as acrylic sheet (e.g. 'Perspex'). Acrylic sheet is usually supplied with a white paper coating lightly adhering to the plastic surfaces. It is not advisable to remove this paper coating because it enables shapes to be clearly marked out with sharp pencils.

Try squares

Try squares for both wood and metalwork are similar in construction in that they consist of a blade held firmly at 90° in a *stock*. However engineer's try squares are of an all-steel construction, whereas woodwork squares are made with a wood stock, with a working edge faced with brass. Blade lengths of from 100 mm to 300 mm are common for woodwork try squares. Engineer's try-square blades are rarely longer than about 150 mm.

If a try square is thought to be 'out of true', then the simple test shown here in a drawing can be conducted. If the lines drawn along A and B do not coincide, then the try square is not true. Careful filing of blade edges may bring the square back to truth.

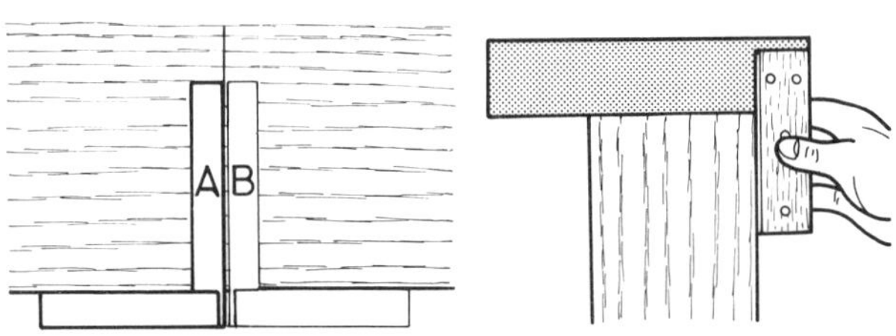

(a) Checking accuracy of a try square and (b) one use of a try square

Pencils, knives and scribers

The shapes required from wood are marked-out in pencil and the lines which are later to be sawn or chiselled are then *cut-in* with a marking knife. Marking and cutting knives are valuable tools for making design models from cardboard, paper and thin plastic sheet.

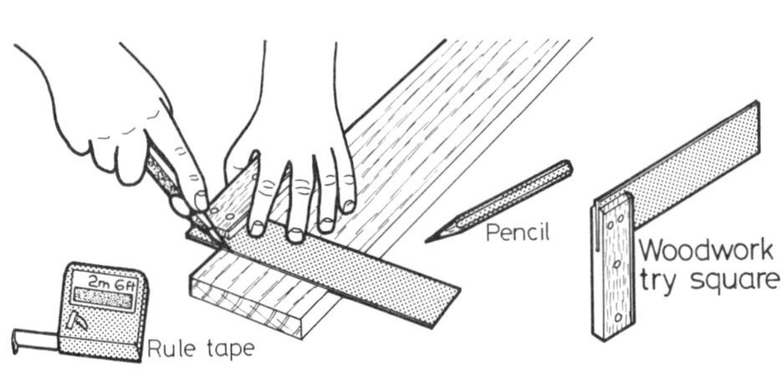

Squaring a line across wood

Most of the marking-out of metals is carried out with *centre punches, dot punches* and *steel scribers* (see pg 13). When marking on the surfaces of metals such as bright mild steel (BMS) a surface coating of 'blue' allows scribed lines to show clearly. On other metals, for example, black mild steel and copper, a surface coating is not necessary.

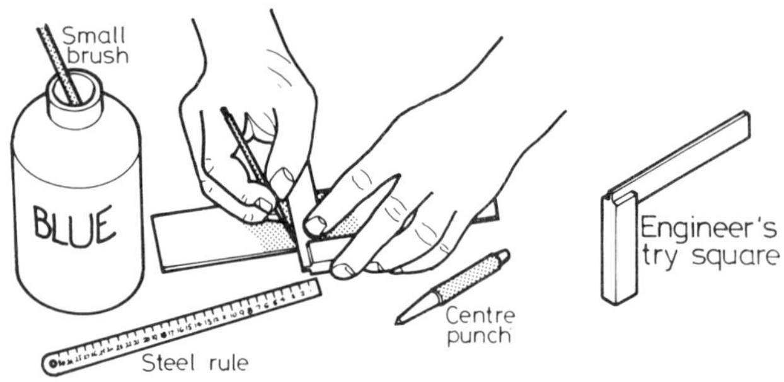

Squaring a line across metal

Rules

300 mm long steel rules are common in all school craft work. Tape rules with tapes 2 or 3 m long, for marking long lengths, can also be used when measuring the shorter lengths for which a 300 mm steel rule is made.

Squaring a line around wood

See Drawings 1 to 4. Note that the stock of the try square must always be placed firmly against either the face side or against the face edge if the lines are to meet at the last corner.

1. Draw a line across the face side with the square held firmly against the face edge.
2. Carry this line across the face edge with the try square held firmly against the face side.

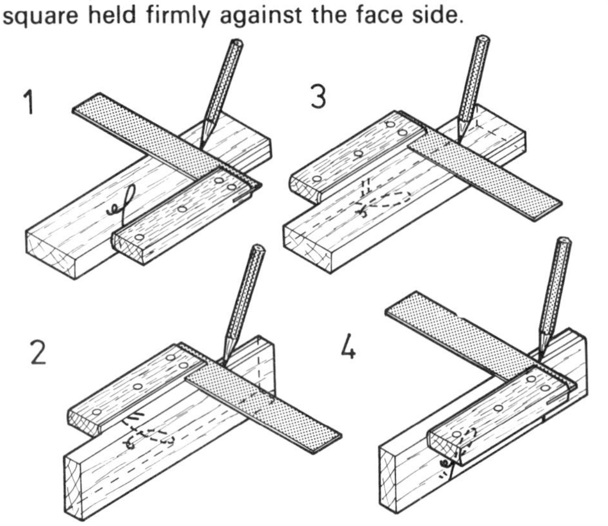

3. Carry the line across the back of the wood ensuring that the square is held firmly against the face edge.
4. Finally, complete the squaring around by carrying the line across the last edge, keeping the try square held firmly against the face side.

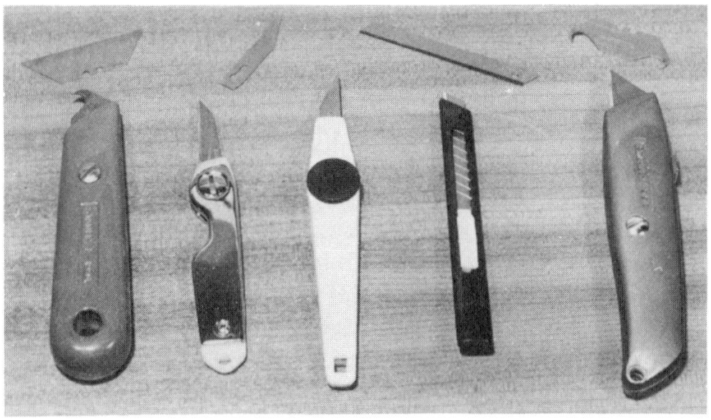

A variety of marking/cutting tools

Marking lines on metals

Dividers

When it is necessary to scribe lines which are parallel to an edge on a piece of metal, *odd-leg dividers* will be found to be suitable. For marking arcs and circles, *spring dividers* can be used. Note that if the centre of a required arc or circle can first be lightly dot-centred, then drawing the arc is made easier because the leg of the dividers will not slip in the dot-centre.

Surface block and surface gauge

A *surface block* is a fine grade cast iron block with its upper surface accurately ground within very fine limits. The ground surface must be kept clean and should be covered when the surface block is not in use. A *surface gauge* can be employed in scribing lines on metal which are parallel to the surface of the surface block on which it is placed. The scriber of the surface gauge can be adjusted for height by means of an adjusting screw or by amending the position of the scriber in its holding device.

Measuring tools for metals

Three measuring tools are illustrated. These are outside calipers, inside spring calipers and a micrometer. *Calipers* are for determining outside or internal dimensions of parts to be measured. When set to the part being measured, the distances between the gaps in the caliper ends are measured on a ruler to determine the required dimension.

A very large range of *micrometers* is made to determine both external and internal dimensions. Micrometers are precision measuring instruments which must be treated with great care. They should always be replaced after use in the boxes in which they were packed and micrometers must not be over-tightened on to, or inside, the parts which are being measured. The micrometer illustrated will measure external sizes from 0 to 50 mm to an accuracy of 0.1 mm.

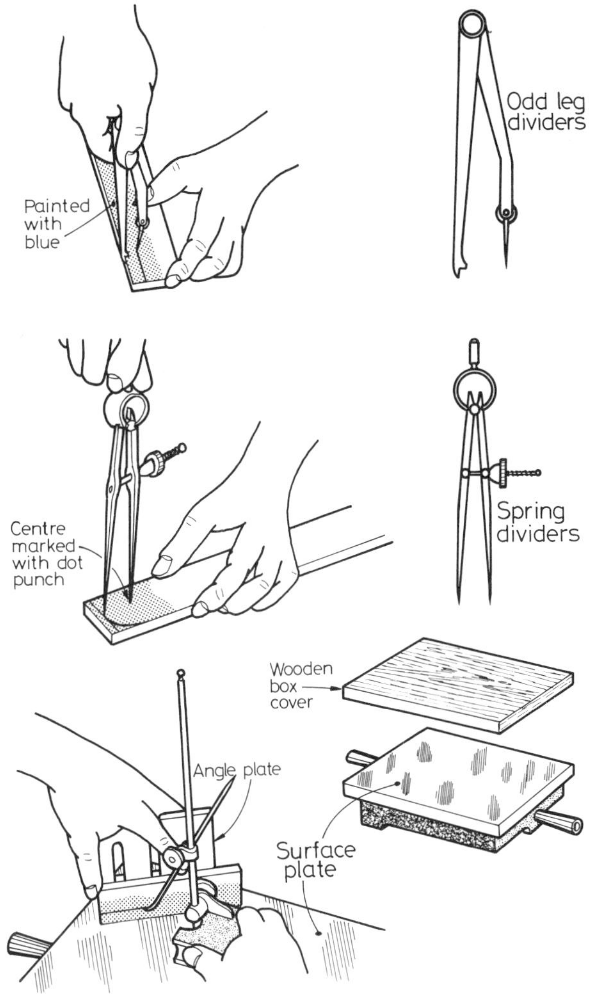

Marking lines on metal

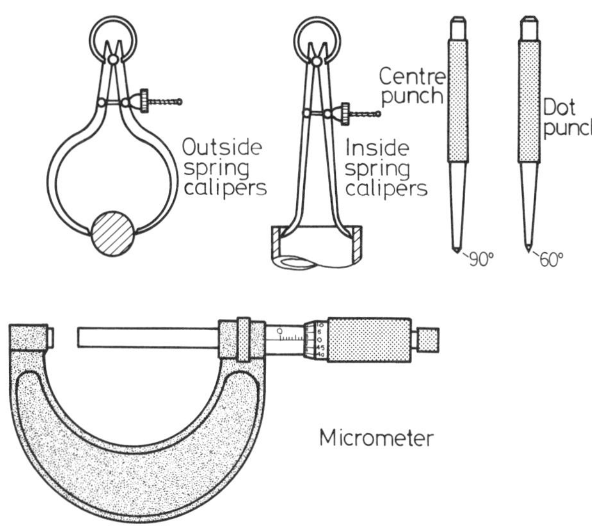

Some marking and measuring tools for metal

Gauges for woodwork

Woodwork *gauges* are for marking or cutting lines parallel to the edges or ends of pieces of wood. They are made from hardwoods and they are fitted with steel spurs or cutters. The stocks of gauges can be moved along their stems and held in position by means of wooden, plastic or brass screws. Some gauges have brass strips screwed into recesses on the working face of their stocks to minimise wear on the wood of the working faces.

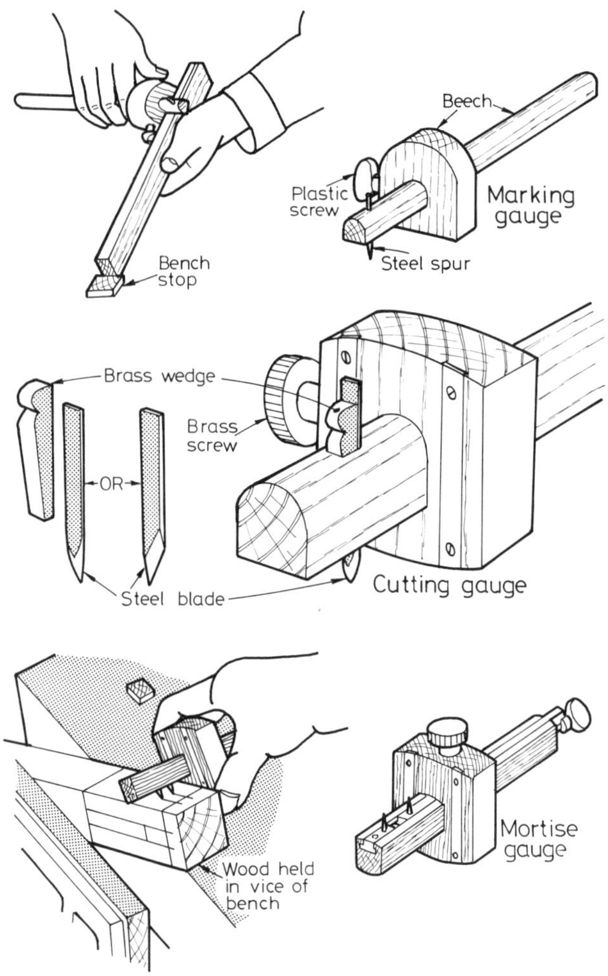

Gauges for woodwork

Marking gauges

These are for marking along the grain lines parallel to an edge or a side, for example: marking lines when planing wood to a required width or thickness; marking lines with the grain when marking out the details of joints.

Cutting gauge

A *cutting gauge* is one in which the steel spur of the marking gauge is replaced by a sharp steel knife set in a rectangular hole and held in place with a brass wedge. The knife can be sharpened in a variety of ways, two of which are illustrated.

A sharp cutting gauge will cut clean lines in wood either parallel to or across the grain. The gauge is for marking the lines of joints such as dovetails when clean-cut lines across the grain and parallel to the ends of the parts making up the joint are required. Wood of up to about 12 mm in thickness can be quickly and easily cut into parallel strips by making deep cuts from an edge with a cutting gauge on both sides of the wood. The parallel strips will then break away cleanly when pressed over the edge of a bench-top under hand pressure.

Mortise gauge

As suggested by its name, this gauge is for marking the parallel sets of lines required for mortise and tenon joints and for bridle joints. Its two steel spurs are set to a required spacing by turning a brass screw at the end of the stem. This operates a slide system to which one of the steel spurs is fixed. When the distance between the spurs has been set, the stock is adjusted to the required distance from the spurs and locked to the stem with a brass screw which also locks the slide system.

Sliding bevel

The blade of a *sliding bevel* can be set at any angle to its stock required and then locked with a locking screw or nut. An example of the use of a sliding bevel is illustrated.

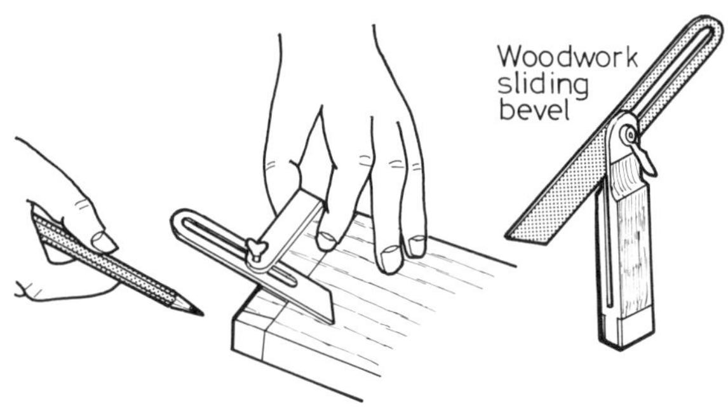

Using a sliding bevel

Face marks

Careful positioning of face side and face edge marks when making and marking constructions in wood will assist in avoiding errors. Note the positions of all face marks in the accompanying drawings.

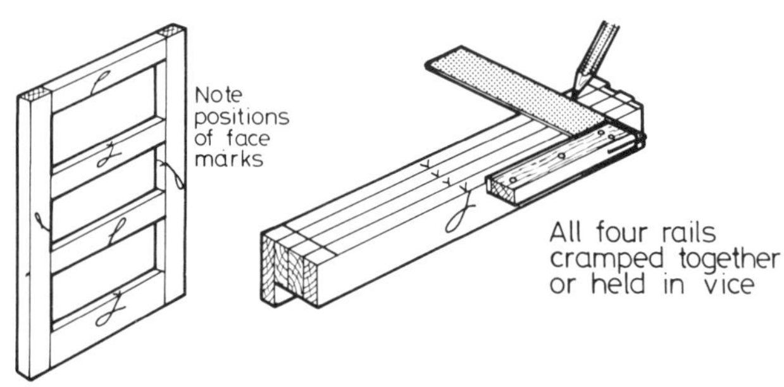

Make good use of face marks

Dot and centre punches

Dot and *centre punches* look similar but their working points are ground at different angles. That of the dot punch is sharpened at 60° and that of the centre punch is at 90°. Both punches are made from carbon tool steel with hardened and tempered points (see page 57 for an explanation of tempering.)

Dot punches have two main uses. The first is for determining precisely the centres of holes which are to be drilled with light dot punch-marks prior to the centres being punch-marked with a centre punch. A second use is for punching light dot-marks along scribed lines. When filing to a scribed line, the line is difficult to see and, in fact, tends to disappear from sight just before the filing reaches the line. The tiny dot-marks stand out clearly as filing nears completion, making the task of filing to a line easier.

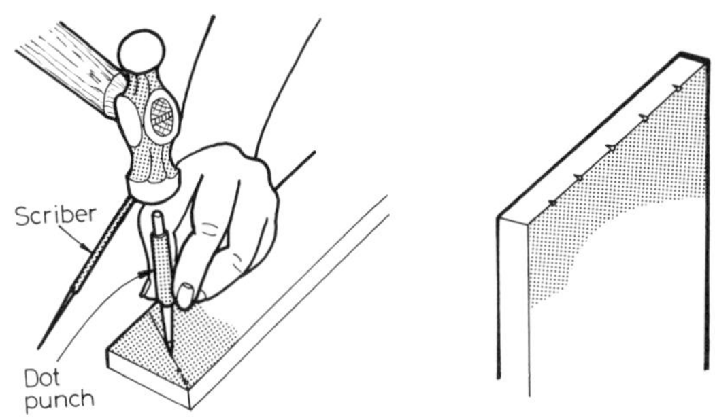

Use for a dot punch

Centre punches are for marking the centres of points at which holes are to be drilled. The indentations of the punch-marks allow a drill to be centred exactly in its required position. The drill is held in the centre punch indentation.

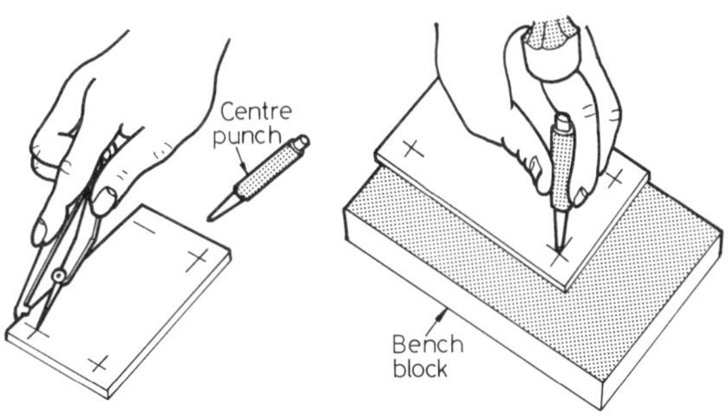

Preparing metal for drilling

Templates

Templates, sometimes spelt *templets*, are for marking out a number of identically shaped parts on the material from which the parts are to be made. Templates can be made from any suitable thin material – sheet metals such as tinplate or aluminium sheet, woods such as plywood or hardboard, plastic sheet, cardboard, paper. The template is accurately cut to the required shape, placed in position on the material and a mark made around the template edges with a pencil or with a scriber. Chalk lines may occasionally be suitable.

Not only do templates enable identically shaped parts to be accurately and speedily marked out, but the spacing between the parts can be easily determined. This makes for economic use of the materials being worked, keeping waste to a minimum.

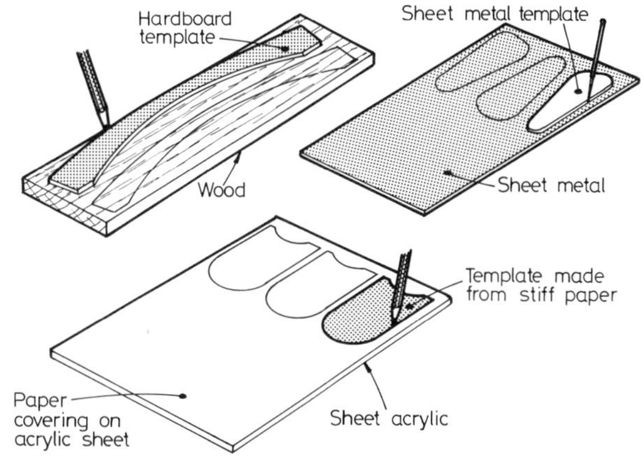

Templates for wood, metal and plastic sheet

Exercises

1. Only those marking tools most frequently used for hand craft work are described in this section. Other tools and equipment will be available in school workshops. Make a list of these. Your list could commence with, for example woodworker's mitre square, engineer's V blocks, . . .
2. On page 12, opposite, one drawing shows the position of face marks on a frame. Make a simple freehand sketch showing the positions of face marks on a wooden box.

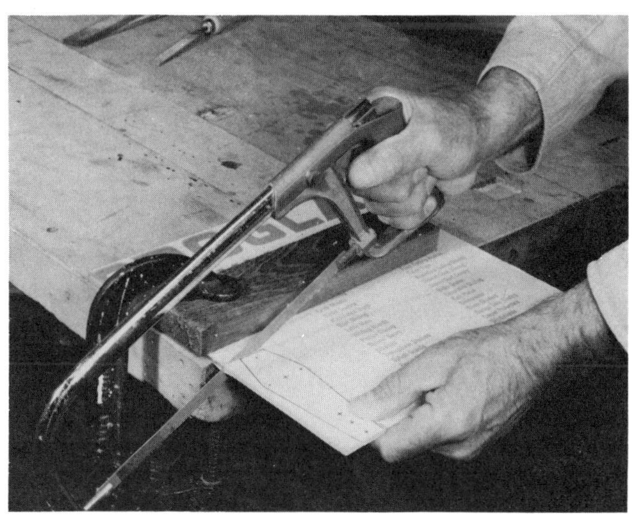

Acrylic sheet sawn with a hacksaw. Note: shape marked out in pencil on paper cover of the plastic.

Saws and sawing

When sawing woods and metals there are a number of saws for specific purposes. When sawing plastic materials by hand most of the common wood or metal saws will be suitable. Thus thermoplastics such as acrylics, vinyls, nylons, polypropylenes etc. can be sawn easily with hacksaws or tenon saws. Some of the harder thermosetting plastics, such as the phenol and melamine formaldehydes found in plastic laminate sheet like 'Formica' are best sawn with hacksaws or with purpose-made tools. Prolonged use of woodwork saws on thermosetting plastics can blunt the teeth of the saw.

Woodwork saws

Handsaws

Handsaws are for the heavy work of sawing wood from boards to the sizes required for the work in hand. They are made with blade lengths of from 500 mm to 650 mm in 50 mm steps, the shorter lengths being more common. There are two main types of handsaw – those with *rip teeth*, and those with *crosscut teeth*. Rip saws are for sawing along the grain of wood, and crosscut saws are for sawing across the grain. The number of teeth per inch (TPI) of a handsaw will often be found stamped on the blade just below the handle. Thus you may find a 6, say, stamped on a rip-saw blade, or a 10, say, on a crosscut blade. Handles are held to the blades with saw handle screws.

Ripping down a board with a handsaw

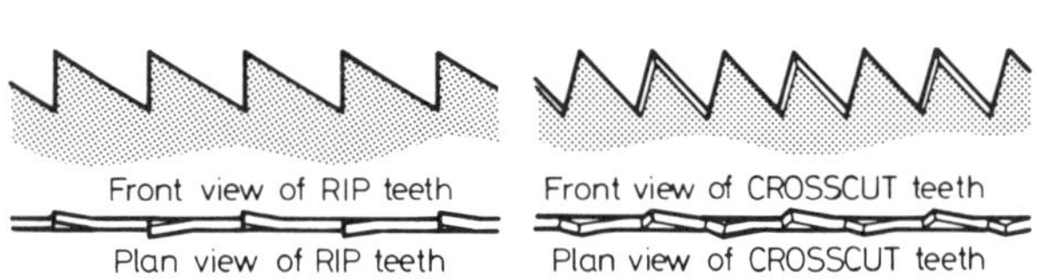

Two types of wood-saw teeth

Backsaws

Backsaws have blades made from thin sheet steel with reinforcing strips of brass (sometimes steel) along their backs. The backs are designed to stiffen the thin blades to prevent them buckling under the pressure of pushing the saw through its cut. Backsaws are made with blade lengths of 350 mm, 300 mm, 250 mm and 200 mm. The three larger sizes are referred to as tenon saws, and the 200 mm length is called a dovetail saw. As these names imply, backsaws are intended for sawing parts of constructional joints such as tenons and dovetails, although their uses extend beyond this. They can be regarded as the general work saws for hand woodwork. Tenon saws usually have 10 or 12 TPI and dovetail saws as many as 20 TPI.

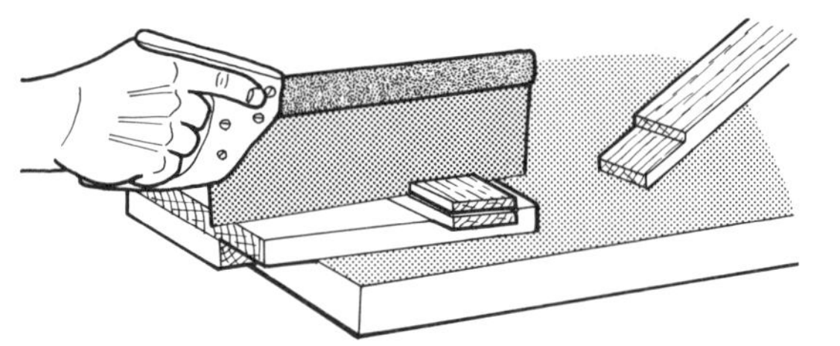

Sawing a halving joint shoulder with a tenon saw

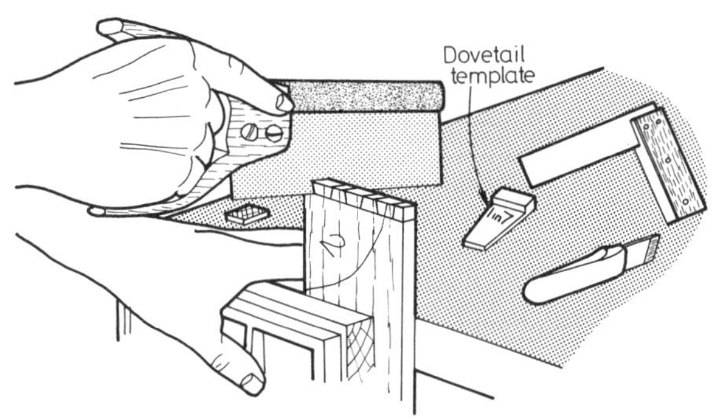

Sawing dovetails with a dovetail saw

Coping saw

Coping saws are for sawing along curved lines in wood up to about 20 mm in thickness. Curved lines in thicker wood can best be sawn with a bow saw (not illustrated) when sawing with hand tools. A coping saw consists of a sprung steel frame between the ends of which 150 mm long disposable thin blades can be tensioned. Blades are thrown away when broken or blunted after continuous use. The blade is held by steel pins each end and tensioned by means of a screw fitted in the handle of the saw frame. The blade can be rotated through 360° to allow it to be worked at any convenient angle along a curved line.

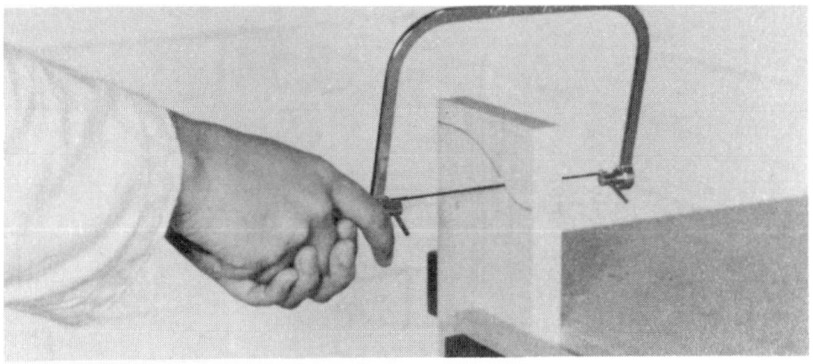

Cutting along a curved line with a coping saw

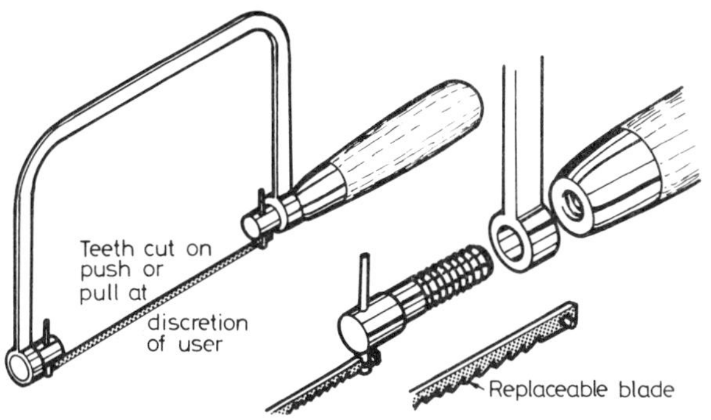

Details of coping saw

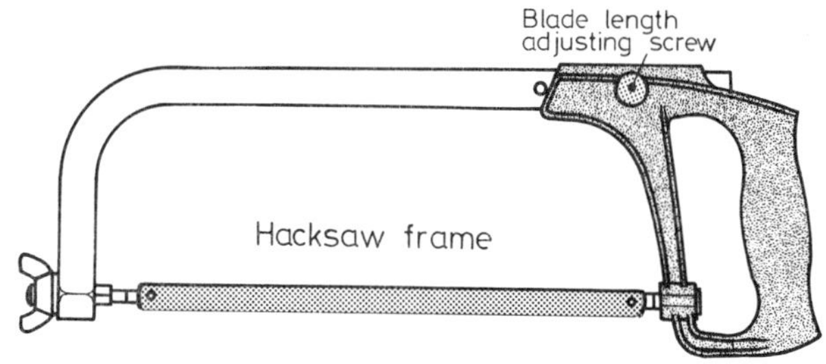

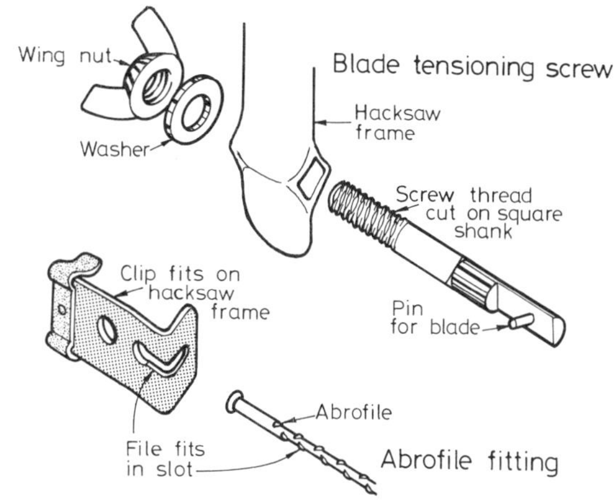

Details of hacksaw and Abrofile

Metalwork saws

Hacksaws

Hacksaws are the general purpose saws for metal-working. A hacksaw consists of a steel frame, adjustable for size to allow blades of either 250 mm or 300 mm length to be fixed in the frame. A tensioning device to hold the blade taut is fitted at one end of the frame. This is a wing nut on a sliding-square screw to which the blade is held on a pin. Blades can be purchased with from 14 to 32 TPI, in flexible or hard patterns, made of tungsten steel or high speed steel. Flexible blades are hardened only along the edge holding the teeth, the remainder of the blade being flexible and thus being less liable to break. Larger teeth are for sawing thick sections in metals such as copper, brass or aluminium. Smaller teeth are for sawing materials of thin sections such as thin sheet or thin-walled pipe. Medium-toothed blades, with say 18 or 24 TPI, are suitable for general bench work. Several precautions, listed below, should be observed when using a hacksaw.

1. Blades should be selected with the correct tooth size and be correctly tensioned in their frames.
2. When sawing thin metal keep the saw angle low.
3. When sawing tubular material use a blade with small teeth.
4. Assume a good working stance. Use slow, steady strokes along a straight line to avoid the blade buckling and breaking.
5. Teeth should point away from the user.

Abrofile

Abrofiles are thin, rod-like blades which, with purpose-made clips, can be fitted in a hacksaw frame. As the name shows, these blades are really files. They are used for cutting along curved lines, even those of very small radius. A typical use for an Abrofile is shown in the photograph. Note that the sheet plastic is held firmly to the bench top with a piece of softwood, itself firmly held under the pressure of a pair of G cramps. This firmness is necessary when using an Abrofile to prevent movement of the plastic, which may cause the file to break as the cut is made.

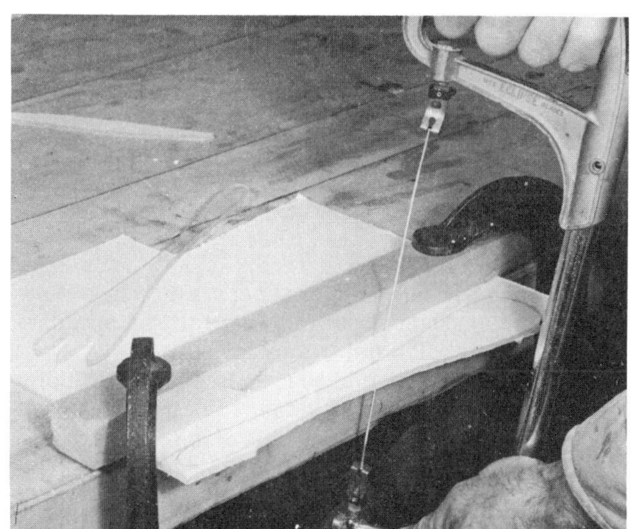

Sawing acrylic sheet with an Abrofile

Junior hacksaw

This is a small hacksaw with a standard 150 mm long blade, which is sprung into a frame made from steel rod. It is a useful little saw for sawing small parts and for items with thin walls, such as thin-walled tubing, where the small teeth are particularly effective. Blades are thrown away when worn out or damaged.

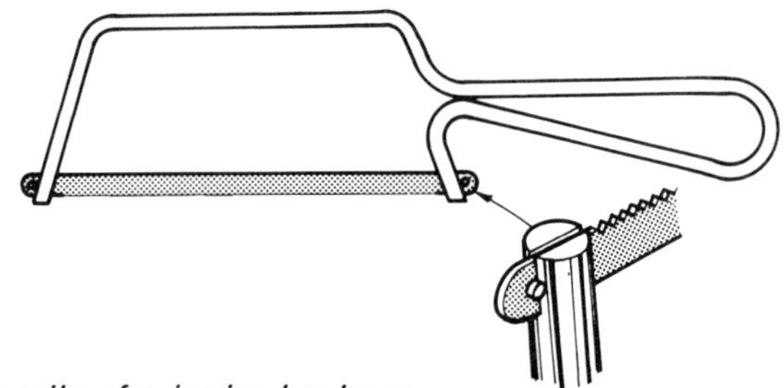

Details of a junior hacksaw

Piercing saw

A *piercing saw* consists of a frame into which a thin, narrow steel blade is tensioned. The saw is for cutting intricate shapes in thin sheet metal. It can equally well be used when cutting shapes from thin plastic materials. The saw is used at a purpose-made sawing board, which is screwed to, or cramped on to, a bench top. The material being worked is held firmly with one hand on to the sawing board, while the other hand works the saw vertically up and down inside a circular slot at the end of a V-cut in the board. The saw cuts on its downwards stroke, when the saw is in tension.

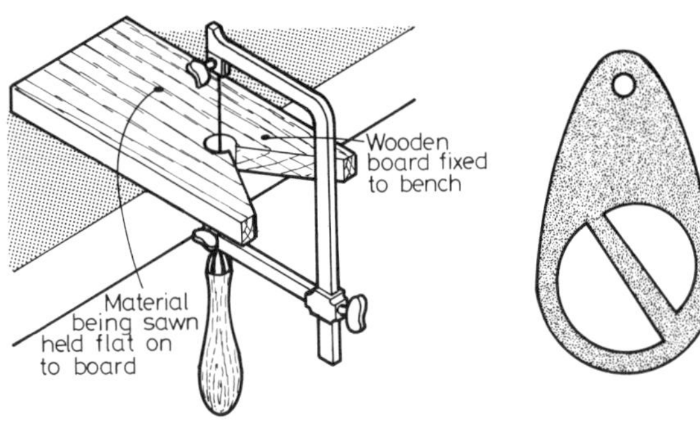

Piercing saw and a pendant formed with a piercing saw

Sawing sheet material

When sawing sheet metal or sheet plastic materials a small-toothed blade should be fitted to the hacksaw and the angle of cutting should be as low as is practicable. These two precautions enable as many saw teeth as possible to be engaged in the saw cut. If the teeth are too large, or the angle of the cut too steep, there is a danger that the saw teeth will jam across the thin sheet causing possible damage to the work or breakage of the saw. Note also that, in the drawing, the saw is shown cutting to the outside of the marked line, so that all of the saw cut is on the waste side of the line.

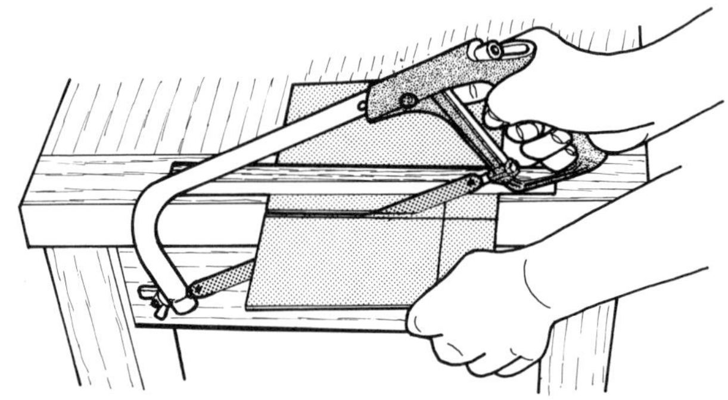

Sawing into the waste

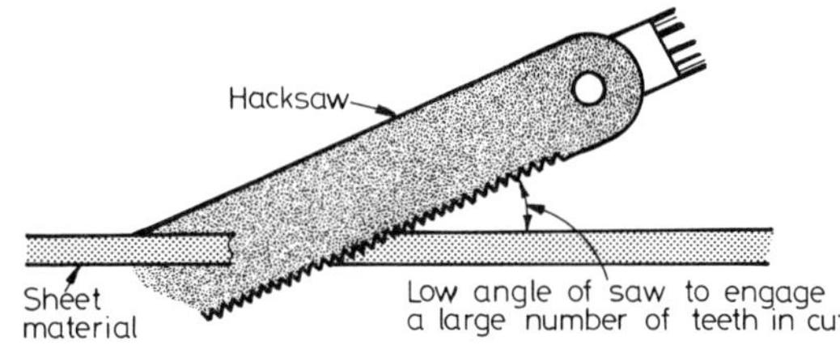

Angle at which thin sheet material should be sawn

Cutting acrylic sheet

Cutters such as that shown in a drawing will make continuous straight cuts along or right across acrylic sheet. Four to six strokes of the cutter will be necessary when cutting sheet up to 4 mm thick, more strokes for thicker sheet. The cutter does not cut right through the sheet but makes a deeply scored line.

A second drawing shows the method by which the piece which has been scored is broken clean. Pressure applied with the palms of the hands and worked from one end of the sheet to the other will break the sheet cleanly along the line. The freshly cut edges can be sanded smooth with fine sandpaper.

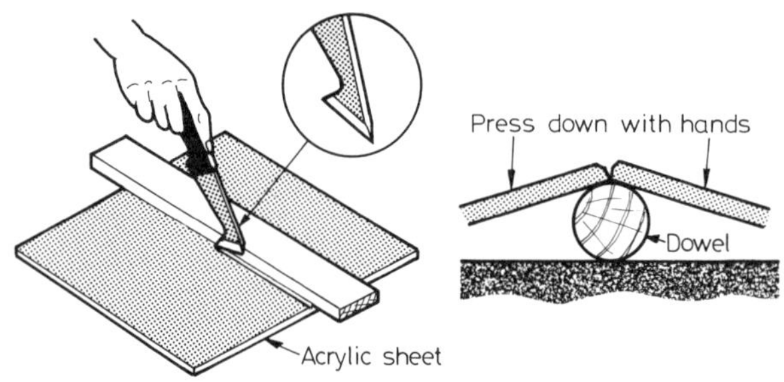

A sheet acrylics cutter and breaking sheet along its cut

General-purpose saw

The photograph on page 17, opposite, shows a *general-purpose saw* being used to saw a length of aluminium tube. The saw has a disposable hard steel blade which is coated with a low friction plastic. This blade is 315 mm long and is designed to allow the saw to cut wood, metals or plastics. The blade can be rotated to several different positions on its handle to allow the saw to be used in awkward positions. Note the engineer's vice on its own board cramped to the bench with Quick-grip G cramps (see also page 6).

Sawing piping with a general-purpose saw

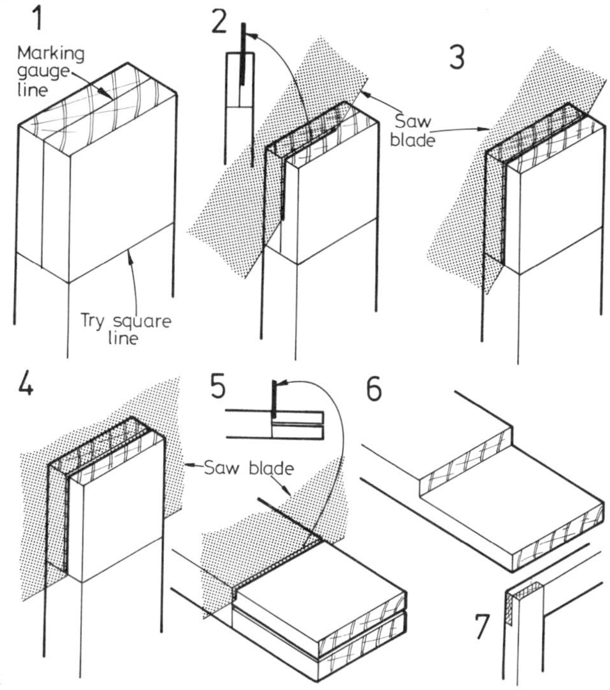

Stages in sawing a halving joint

Sawing a halving joint

The six stages involved in sawing the two parts of a halving joint for a wooden frame are shown here to demonstrate the importance of *sawing into the waste.* The most important rule involved in all forms of sawing is to ensure that the saw is cutting accurately on the waste side of the lines marked on the materials. Unless this rule is strictly observed, either the parts of a construction will be too small and thus not fit together, or parts intended to be made to particular dimensions will be too small. Thus, when sawing, always saw into the waste. This has already been mentioned in connection with sawing sheet materials. See page 16, opposite.

The stages in making the halving joint are as follows

1. Mark the parts of the joint.
2. Saw at an angle, along the vertical line taking care that the saw cut is made into the waste wood.
3. Turn the wood around and make a similar cut from the other edge of the wood.
4. Complete the saw cut to its full depth, the saw being guided in the two sloping cuts already made.
5. Saw the shoulder of the halving, again ensuring that the saw cut is all into the waste of the wood.
6. The completed cuts.
7. The two parts of the halving assembled together.

Sawing a large circular hole in wood

When a large circular hole is required in a piece of wood, the procedure is as follows.

1. Mark the circle of the hole with a compass and pencil. Bore a hole of, say, 6 mm diameter, inside the circle – that is into the 'waste side' of the circle line. Release the blade of a coping saw from its frame and insert the blade through the hole. Replace the blade in its frame slots and tighten the handle of the frame to tension the saw.
2. Now the hole can be sawn. Ensure that the saw cut is made on the inside (the waste side) of the circle, otherwise the hole will be too large. As the sawing proceeds it may be necessary to rotate the saw in the frame to a more convenient angle.

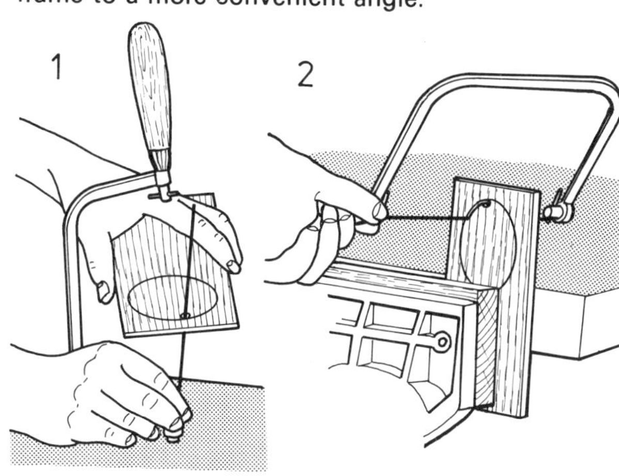

Stages in sawing a hole with a coping saw

Planes and files, planing and filing

Planes are for forming wood to size and shape. Files are mainly for forming metals and plastics to size and shape.

Planes

Wood planes

Practically all modern wood planes are made with bodies of fine cast iron to which cutters and their adjusting mechanisms are fitted. Planes with wooden bodies are, however, occasionally of value.

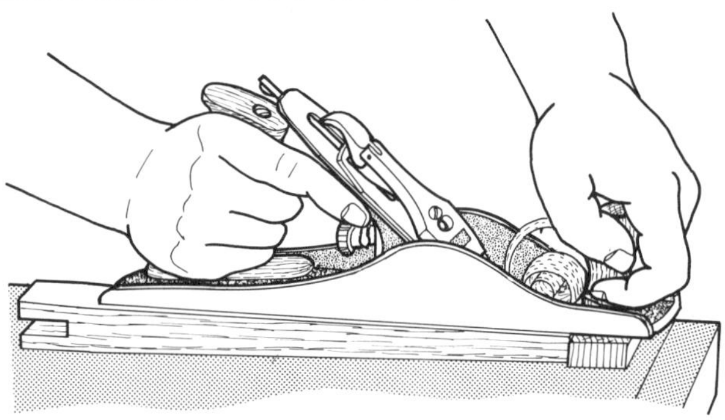

Action of a well-sharpened plane

Bench planes

Two bench planes are in common use: A *jack plane* (300 to 340 mm long) and a *smoothing plane* (245 mm long), both of which are usually fitted with 50 mm wide cutters. Other sizes of this pattern of plane are also manufactured.

Jack planes are for the heavy planing involved in reducing pieces of wood to their final dimensions of width and thickness, whereas smoothing planes are for producing smooth surfaces prior to sanding. The cutting action of all bench planes is the same. A cap iron is screwed to the cutter to help to avoid tearing as the wood is planed and to deflect shavings out of the plane mouth. The cutter/cap iron assembly is held at 45° to the sole by means of a lever cap and a holding screw. (The lever cap holds the cutter firmly in position under the pressure of a cam lever.) The cutter can be made to make heavy or fine shavings with the aid of an adjust nut and a lateral adjusting lever which control the position of the cutter relative to the plane's sole.

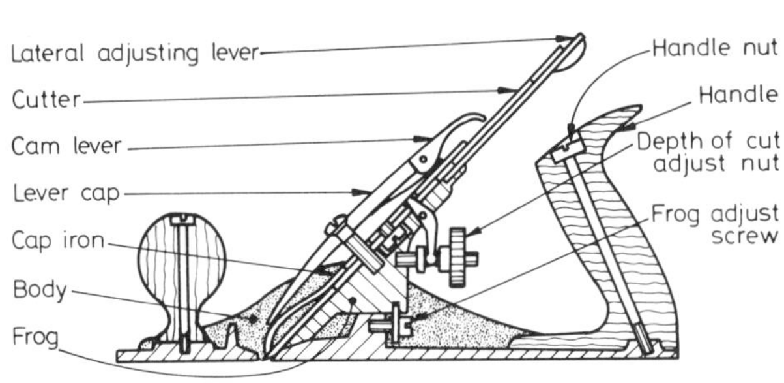

Details of a bench plane

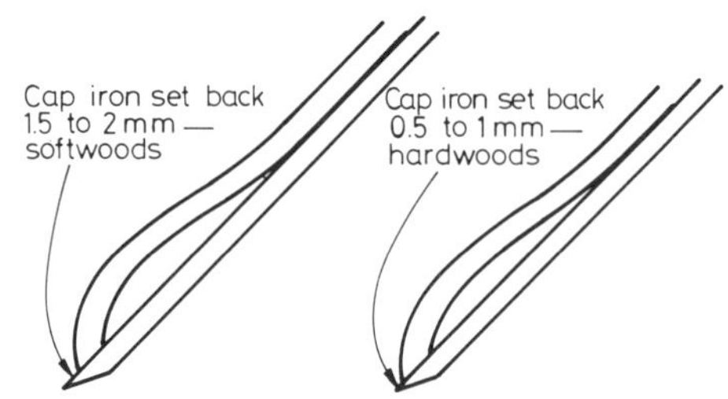

Details of a bench plane cap iron

Planing wood to size

Six drawings show the stages in planing wood to produce a piece of given width and thickness.

1. Saw the unplaned piece from a board – 13 mm longer, 6 mm wider, and 3 mm thicker than the required finished dimensions.
2. Plane the face side and the face edge. These two surfaces should finish flat, straight and square with each other and should be tested for straightness and for squareness with the edge of a rule and with a try square. Mark the face side and the face edge with pencilled face marks.
3. Set a marking gauge to the required finished width and gauge a line on the face side with the stock of the gauge running along the face edge.
4. Plane the second edge of the wood down to the gauge line, to produce the required width.
5. Set a marking gauge to the required thickness and gauge lines on both planed edges with the stock of the gauge running along the face side.
6. Plane the rear of the wood to the two gauge lines to produce the required thickness then test for flatness with a rule or with the blade of a try square.

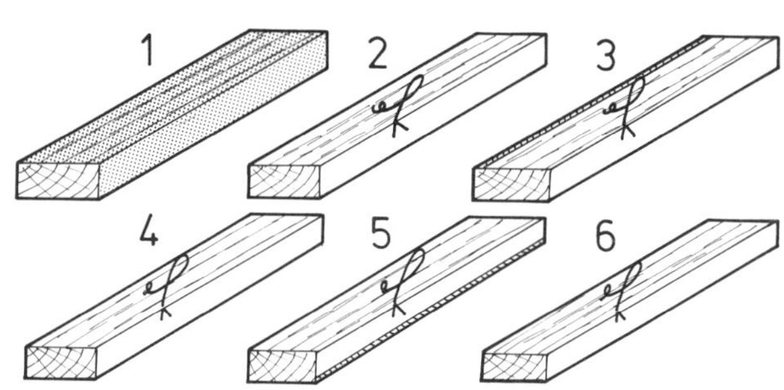

Planing all round a piece of wood to correct sizes

Planing end grain

When planing the end of wood, the wood may split at the far corner as the plane completes its cut. This can be minimised by using a shooting board (page 7) or either of the two methods shown in the drawings below. Chamfering a corner has the disadvantage that a small chamfer is left on the wood after its end has been planed.

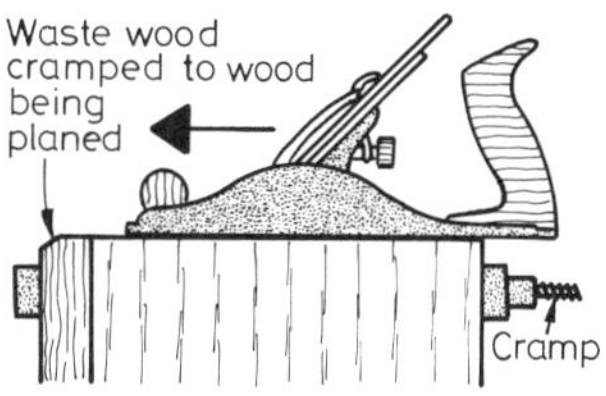

Planing end grain

Smoothing plane, rebate plane, shoulder plane, combination plane and block plane

Other wood planes

Block plane

This is a short plane, about 180 mm long, which can be used in one hand alone. Its cutter has no cap iron and is held at 20° to the sole by a steel cap and holding screw. This plane is for smoothing the surface of wood likely to tear during planing.

Block plane

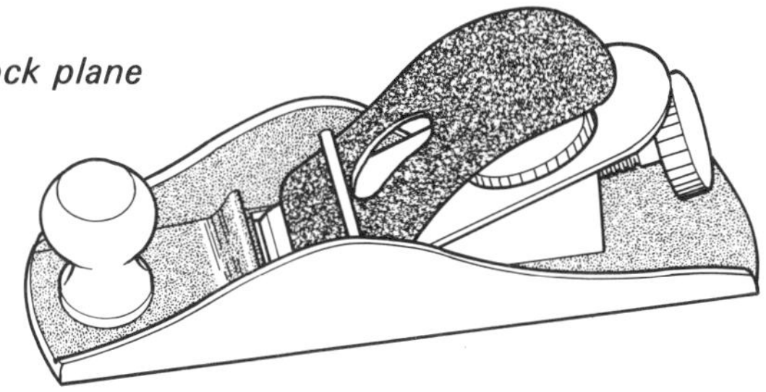

Rebate plane

This is for planing rebates. This plane is equipped with an adjustable depth stop and adjustable fence which enable the tool to be pre-set for planing a rebate to the required width and depth.

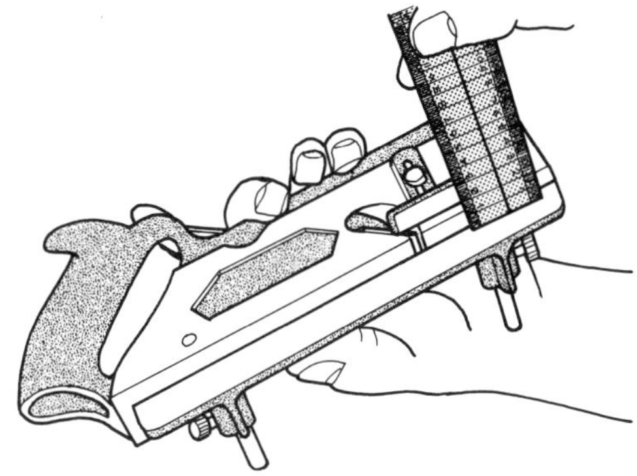

Setting the fence of a rebate plane

Combination plane

This tool has cutters of a variety of shapes and sizes, in order to work widths of groove, tongue and mouldings from 3 mm to 13 mm. A depth stop and fences allow adjustment of the position and depth of the cutter.

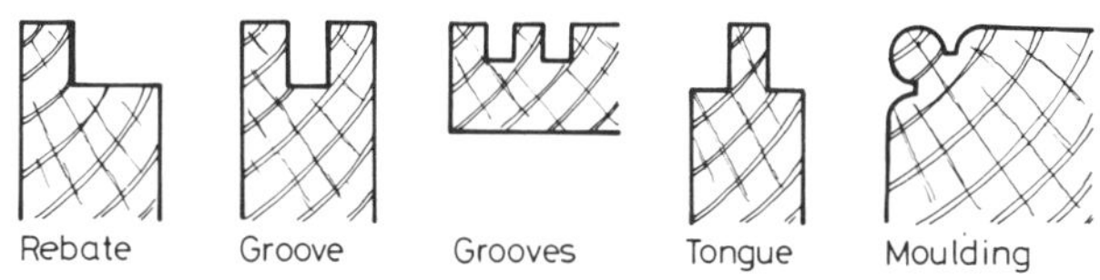

Types of section which can be planed with a combination plane

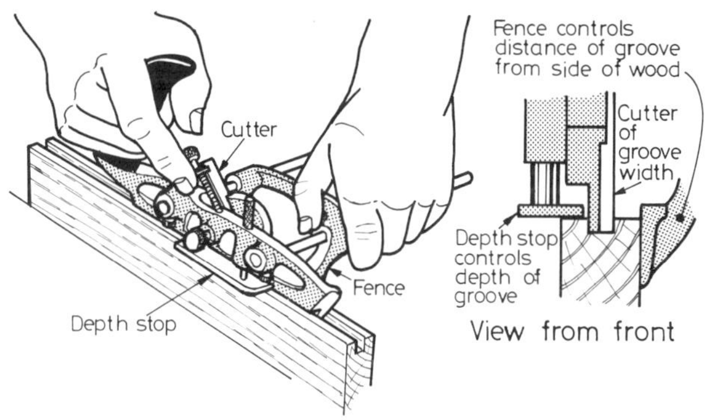

Cutting a groove with a combination plane

Router plane

This is for planing the bases of grooves cut across the grain of wood, to ensure that the bases are parallel to the wood surfaces. It can also be used to produce a flat surface at the rear of a recessed carving. The cutter can be set to protrude below the sole of the plane at the depth required.

Cutting a housing groove with a router plane

Files

General purpose files for metalworking are known as *engineer's files*. These are made in a huge range of lengths, shape and coarseness of cut. Another group of files for fine work is *needle files*, made in a more limited range. Engineer's files are classified by length, shape and grade of cut.

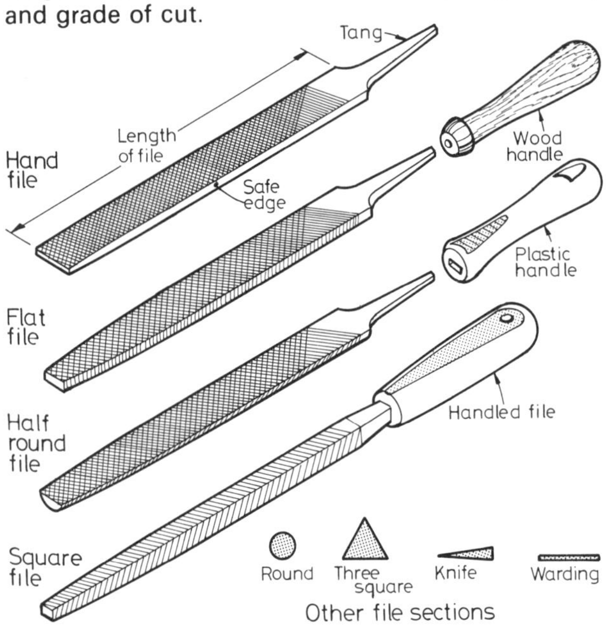

Some engineer's files

Length

An engineer's file is measured from the shoulder above the tang to the end. (Note that needle files are measured from end to end.) Engineer's files are manufactured in lengths of from 100 mm to 350 mm in 50 mm steps. Lengths of files common in schools are 200 mm and 250 mm, although 100 mm and 150 mm long files will occasionally be of value.

Shape

Files are made in a wide variety of shapes as can be seen by reference to the accompanying drawings. Shapes other than those illustrated are also manufactured, as reference to a file maker's catalogue will show. Note that a hand file, while rectangular in section, tapers from end to end in thickness whereas a flat file, while of constant thickness, tapers sideways. Note also that hand files are made with a safe-edge, which is of value when filing against a shoulder of which one part must not be filed.

Cut

Files are made in five grades of cut – *rough, bastard, second, smooth* and *dead smooth*. Of these, bastard, second and smooth-cut files are commonly found in school work. Other special cuts of files are made, in particular for filing brass. It is, however, unlikely that the pupil or student will find such special cuts of much value to their work. Files are commonly *double-cut*, in that two crossing lines of teeth are cut on file faces. *Single-cut* files are also made for light finishing cuts. The cut varies in size with the length of a file. Thus the teeth of a 100 mm second-cut file will be finer than those of a 350 mm second-cut file. Bastard files are for coarse work, second-cut for general work and smooth files for fine cuts prior to obtaining a finished surface.

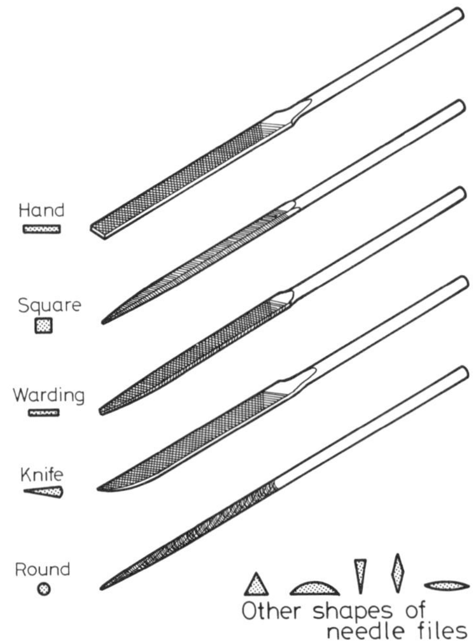

A selection of needle files

Needle files

Needle files are made with round handles as part of the file. These handles allow accurate manipulation of the files during fine precision work. Needle files are made in two lengths, 140 mm and 160 mm (these lengths include the handle). All needle files have smooth-cut teeth.

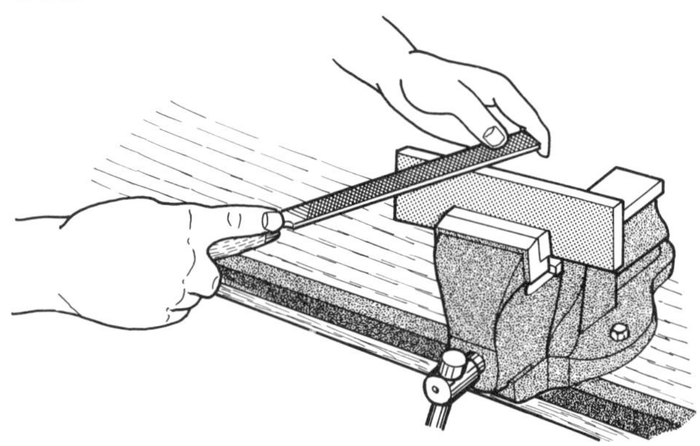

Correct position for filing an edge

Filing

Files are expensive tools because, unlike other cutting tools used in craftwork, they cannot be sharpened in the workshop when blunt. Because of this, files should be treated with extra care. The following rules should be observed.

1. Files must be stored in such a manner that the file teeth of one cannot come into contact with the teeth of another.
2. Engineer's files must never be used without a handle. Accurate filing can be achieved only if a handle is on the file tang. If a handle does not cover the tang, the tang can cause injury to one's hands.
3. Files are brittle and hard and will break if used as levers.
4. Keep the teeth of a file open by cleaning with a file card as necessary (see also page 59).
5. Blackboard chalk rubbed into the teeth of a smooth-cut file will assist in keeping the teeth free of debris and so allow smooth filing without pinning.
6. Choose a file of the correct cut for the work in hand. Bastard-cut for heavy removal of material; second-cut for general filing to achieve a fair finish; smooth-cut for fine finishing.

Rules when filing

Much practice is required to achieve a good standard of accurate filing. Some rules which will assist in reaching a good standard are as follows.

1. Stand facing the work with the feet wide and the body comfortably, yet firmly, positioned above the feet.
2. In general the file should be held between both hands.
3. Set the work in the vice as low as is practicable to prevent it chattering as filing proceeds.
4. Make long, steady strokes, using as much of the length of the file as possible. Rapid strokes will not achieve rapid or accurate filing.
5. Apply sufficient, but not too much, pressure downwards on the file to ensure that the file is cutting well.
6. Hold the file at an angle to the length of the work and, as far as possible, in each stroke work from one end of the work to the other.
7. Begin with a bastard-cut to remove excess waste, then finish with a second-cut. If a fine finish is required then go on to use a smooth-cut.

Filing curved edges

When filing a curved edge, use the best sectional shape of file available and make each file stroke in such a manner that it commences at one end of the curve and during the stroke moves to the other end. If a long curve is being filed, this may have to be shaped in several stages, but the same principle should be applied.

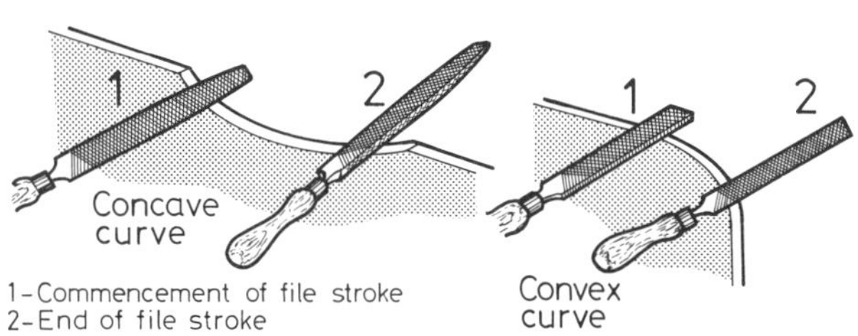

Positions of files when shaping curved edges

Draw-filing

Draw-filing is a process suitable for obtaining a very smooth surface by filing. A smooth-cut file is held between the fingers of both hands and worked backwards and forwards along the surface being smoothed. Draw-filing removes very little material and achieves a fine surface which can be finished with emery cloth. If a surface of a small piece is to be draw-filed it can be held between pins on a piece of scrap wood.

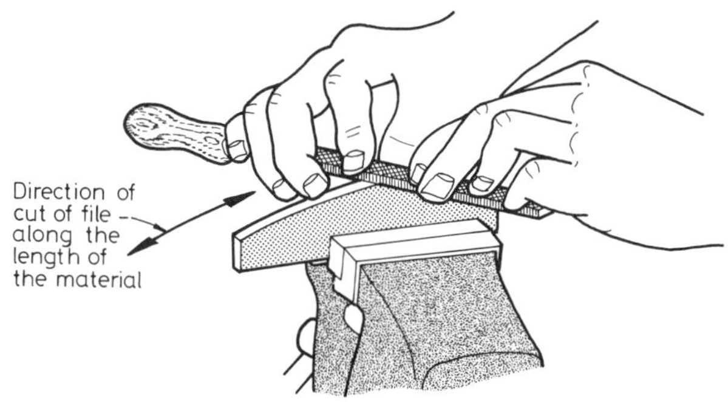

Draw-filing an edge

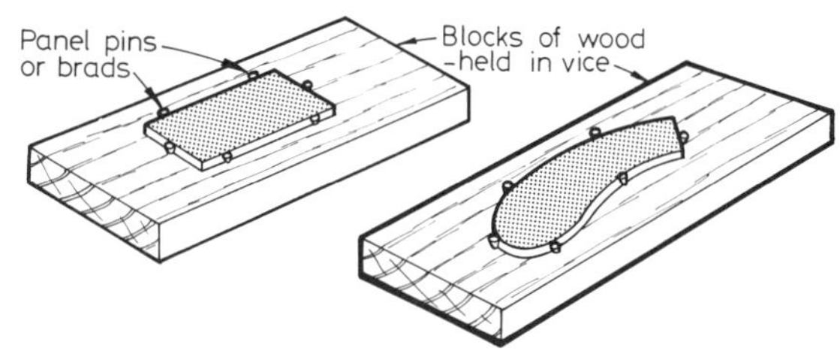

Pieces of sheet metal attached to scraps of wood for draw-filing and polishing

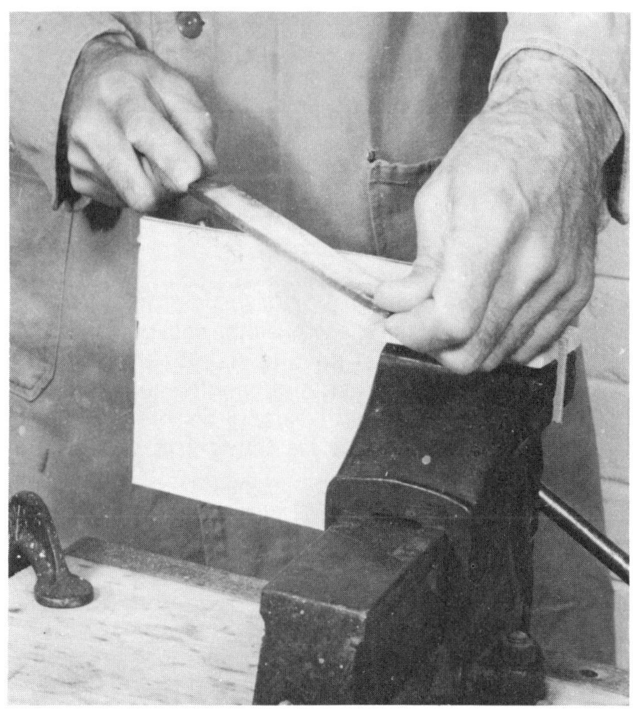

Filing the edges of acrylic sheet

Tools for shaping wood

Spokeshaves

Spokeshaves are similar to planes in their action. A 50 mm wide cutter is held in the malleable iron body of the tool by means of a fixed screw and a tightening screw. The depth of cut is controlled by means of a pair of adjusting nuts on screws, which fit into slots in the cutter. Handles each side of the body allow the tool to be firmly held. Two standard forms of spokeshave are made – one with a flat bottom, the other with a rounded bottom. The first is for the shaping of outer, convex curves; the second for shaping inner, concave curves.

When in use, a spokeshave will cut more easily and freely if the tool is held in such a manner that the cutter is at an angle to the work, rather than square with it. If held square to an edge, not only is more pressure needed to make the tool cut, but there is a tendency for the tool to chatter and not to remove good shavings.

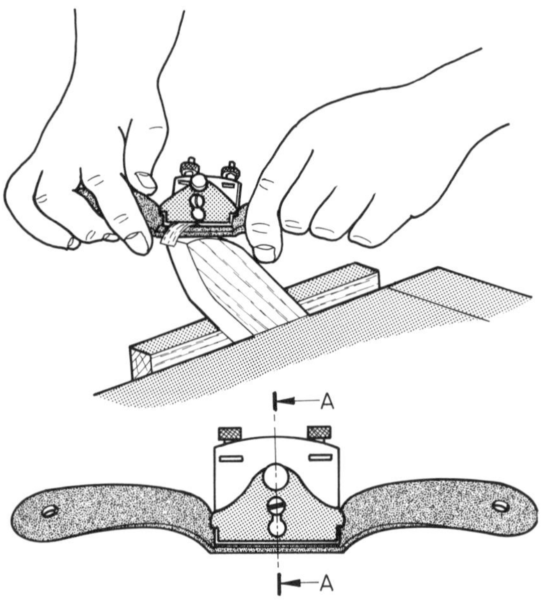

Shaping a curve with a spokeshave

A note on using planes and spokeshaves

A light dressing of candle wax on the sole of planes and spokeshaves will assist them to slide freely along the wood being worked. A stub of candle can be kept for this purpose and just rubbed along the sole of the tools when the metal bottoms are tending to stick to the wood.

Details of spokeshaves

Wood files and rasps

Bastard-cut and second-cut engineer's files are suitable for filing the shaped edges of wood, although purpose-made half-round *wood files* are manufactured. Hand and half-round *wood rasps* are also made for the rapid removal of waste prior to filing. The rasp and file shown in the photograph are *handled*, which means that they are made with a plastic handle permanently fitted to the tangs of the tools.

When shaping the edges of wood with a rasp or file the stance and action of working are exactly the same as when filing metal. Thus the same rules of filing should be observed.

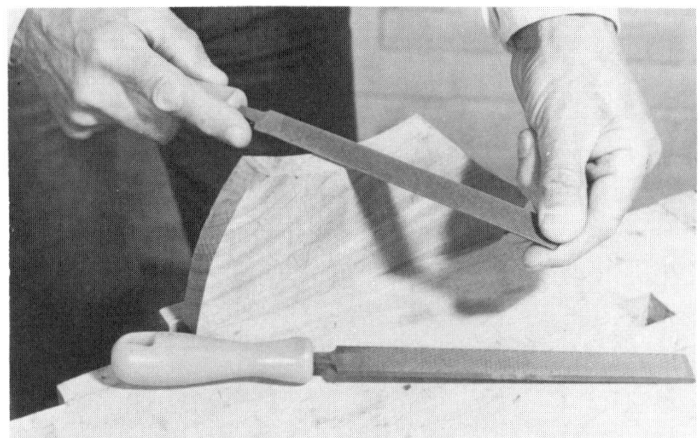

Filing curves in wood with wood rasp and wood file

Surform tools

These are for the rapid shaping of woods, some metals and some plastics. Each Surform tool consists of a body into which a flexible blade is fitted. The blades contain teeth which are shaped so as to allow shavings to pass right through the blade. When a blade becomes worn or damaged it can be removed from the tool body and a replacement blade fitted in its place. Blades are made in three grades of cut: *standard-cut* blades are for shaping softwoods and plastics such as nylon; *fine-cut* blades are for finishing softwoods, for shaping hardwoods, for plastics such as acrylics and for soft metals such as aluminium; blades with grades of special-cut are for continuous use when cutting woods, metals such as aluminium, copper and brass and the edges of plastic laminate sheet.

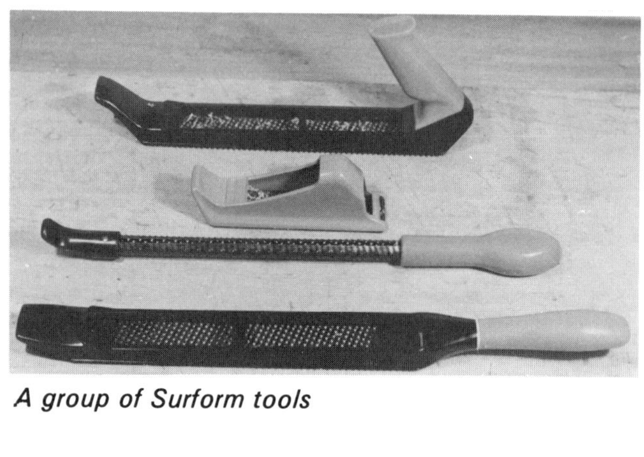

A group of Surform tools

Chisels and chiselling

Wood chisels

As can be seen in the photograph on page 24, overleaf, *wood chisels* are made in a range of different forms and sizes. In school and college work three types of wood chisel are common – *firmer chisels*, either *square* or *bevel-edged; mortise chisels; heavy-duty chisels.*

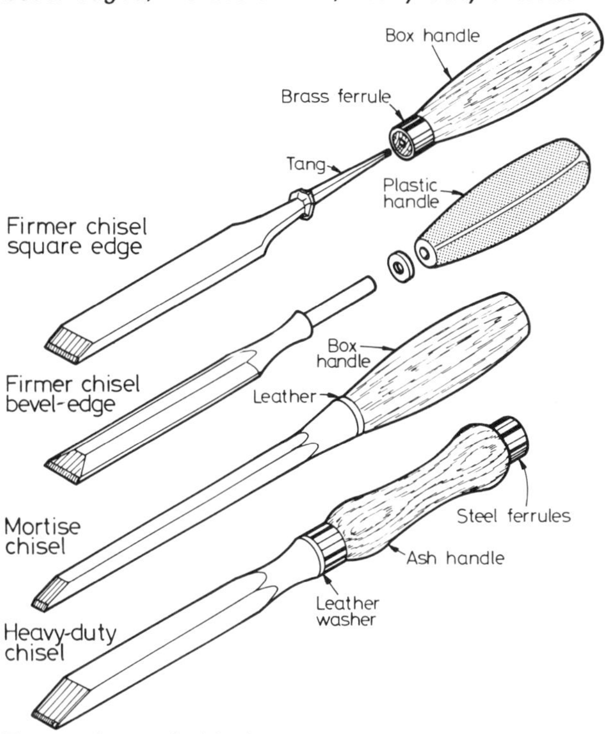

Types of wood chisels

Firmer chisels

Firmer chisels are for forming wood to the various shapes required in designs (the name *'firmer'* may come from the word *former*). A set of five firmer chisels of blade widths 3 mm, 6 mm, 13 mm, 19 mm and 25 mm is sufficient to deal with all the work suitable for these tools. The bevels of bevel-edged firmer chisels allow the chisel to fit into angled corners, as, for example, in chiselling waste from between dovetails. This may make these chisels more useful than square-edged chisels. Handles may be of wood (ash, box, hickory) or of plastic such as cellulose acetate butyrate (CAB).

Mortise chisels

These chisels must withstand the strain of mallet blows on their handles, so the blades are heavier and shorter than those of firmer chisels. A leather or nylon washer is also fitted between the handle and blade to assist shock absorbance. The blades are available in 3 mm, 6 mm, 8 mm, 10 mm and 13 mm widths, and the handles are often of boxwood.

Heavy duty chisels

The blades of this tool are stout and heavy compared with other wood chisels. These chisels are made for very heavy work involving driving the tool with hammer blows. However do not use a mallet on heavy duty chisels as the steel-ringed handle will rapidly damage the mallet faces. Mallets should also not be used on plastic-handled chisels. Both steel-ring and plastic-handled chisels are best driven when necessary with hammer blows.

Using wood chisels

Four drawings show methods of using wood chisels. Note that in each of the four examples both hands are engaged in either holding the tools or holding the wood. Chisels must be sharp if they are to function well but sharp tools can cause severe injuries if handled wrongly or carelessly. If both hands are engaged in holding the tools or the work in such a manner that the cutting edge of the chisel is away from the hands or body, then accidents causing injuries will not arise.

Paring with a firmer chisel (left) and chopping a mortise with a mortise chisel (right)

Cutting a groove with a firmer chisel (left) and paring a curve (right)

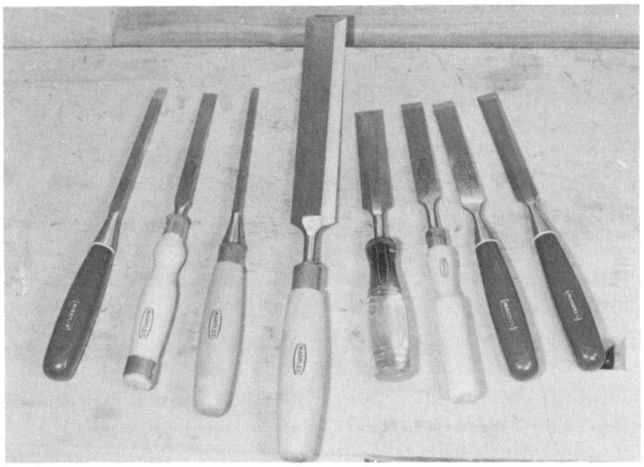

A selection of wood chisels

Gouges

Bench gouges

Bench gouges are made in the same style as firmer chisels, with a handle and ferrule (the metal ring at the blade end of the handle), but the blade, is curved across its width instead of being flat. Gouges are made in a number of different widths and with different radius curves.

There are two types of bench gouge: those sharpened on the outside of the curve of the blade are *firmer gouges*; those sharpened on the inside of the curve of the blade are *scribing gouges.* Bench gouges are used in much the same manner as firmer chisels except that they are for curved work.

Bench and carving gouges

Carving gouges

Carving gouges are lighter in weight and construction than bench gouges and they are made in hundreds of different widths and different blade curvatures. Only a small selection from this large number is illustrated here. When working with carving tools always ensure that the wood being worked is firmly held either in a vice or on to a bench top.

Cold chisels for metals

Some chisels are used when working metals which are red hot, as when cutting hot metal from a forge at an anvil with a hot sett or hot chisel. The term 'cold' refers to the fact that the *metal* being cut is cold. A number of different forms of cold chisel are made but by far the most frequently used is the *flat cold chisel.* Among other cold chisels which are made, are *cross-cut, round-nosed* and *diamond-pointed* (see page 25). School and college work only rarely requires other than flat cold chisels.

Cold chisels are made from carbon tool-steel or from nickel chromium tool-steels of hexagonal section in a number of different sizes and lengths. The flat cold chisels most often used have 16 mm or 19 mm wide cutting edges, and are 150 mm or 200 mm in length but narrower, wider and longer chisels are also made. The cutting edges of all cold chisels are normally ground at 60°.

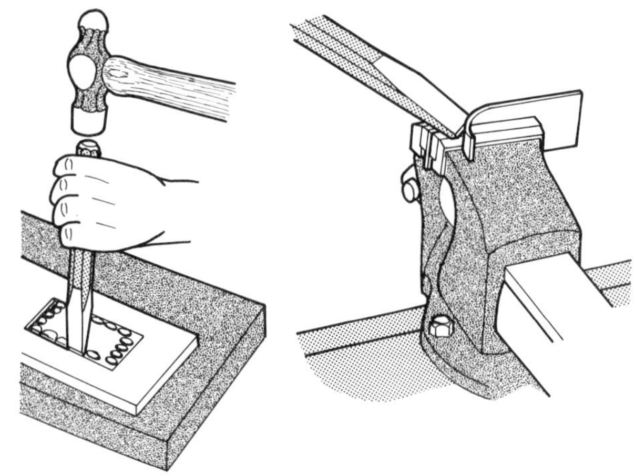

Methods of cutting with a flat cold chisel

Using flat cold chisels

Two illustrations indicate methods of working with flat cold chisels. These tools are made to cut under blows from an engineer's hammer but a hammer of sufficient head weight is essential as too light a hammer tends to make the task of cutting metal with a cold chisel somewhat tedious. Note the following points.

1. When a rectangular hole is to be cut in metal, a series of drill holes bored in the waste of the metal will make the task of chiselling out the hole much easier.
2. The chisel cuts must be made to the waste side of the lines, and excess material remaining after chiselling can be removed by filing.
3. A heavy chipping block under the metal will prevent damage to the bench top.
4. Waste from sheet metal can be removed by placing the metal in an engineer's vice and slicing off waste as shown in the illustration.

Other cold chisels

Cross-cut and round-nosed cold chisels, together with examples of the types of cut made by these two chisels, are shown in drawings. Cross-cut chisels are made with blade cutting-widths of 3 mm, 6 mm and 9 mm, the blades being wider than the handles to give added strength. Round-nosed chisels are made to cut channels of 2 mm, 5 mm and 8 mm radius.

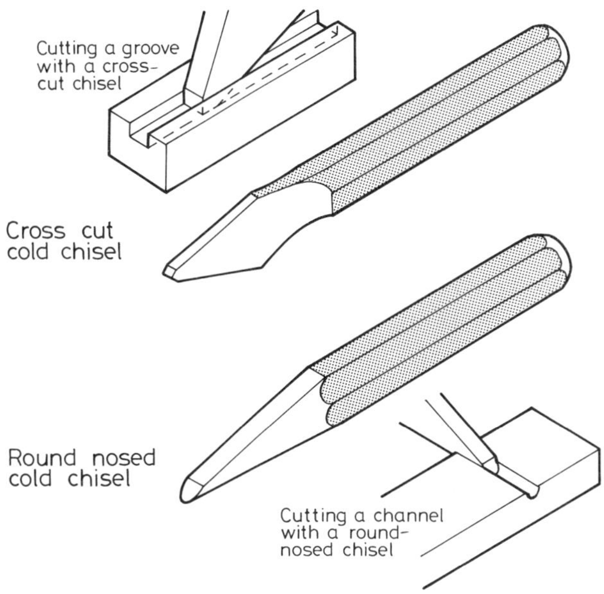

Types of cold chisel

Hand turning-tools for wood

The shaping of wood on a lathe is carried out with a number of gouges and chisels all specifically designed for the purpose. The more important of these tools are shown in drawings. Reference to turning-tools will be made again on page 52. Hand turning-tools for wood are longer than bench chisels and gouges and generally made of stouter proportions being generally about 600 mm long overall. This length and stoutness enables the operator to hold the tools very firmly in both hands, tucking the handle into his or her side. This is necessary to ensure that tools do not move unnecessarily under the pressure and strain exerted on the rotating wood by the lathe.

Three types of hand turning-tool are mainly used – gouges, chisels and scrapers.

Turning gouges are for the heavy cutting involved in reducing wood to a cylindrical form and then turning it to roughly its required shape. *Turning chisels*, in the hands of experienced turners, are then used to complete the shaping to an accurate and smooth finish. However to produce such accuracy of finish requires considerable experience and practice and the school or college student will frequently find it necessary to achieve a desired shape with a number of different *scraper chisels* which are considerably easier to manipulate.

Scraper chisels do not produce such a smooth surface as turning chisels, except when working very hard fine-grained woods such as box or rosewood. Thus, when scrapers are used a fair amount of sanding with sandpapers will be requied to achieve a satisfactory smooth finish. Note that the cutting action of gouges and chisels is a true cutting one, achieved by placing the cutting angle of the tool at a tangent to the rotating wood over and above the lathe centres. The action of scraper chisels is a scraping one, the tools being held at right angles to the face of the wood as it rotates and in line with the lathe centres.

The angle of sharpening of turning gouges and chisels is the same as for bench chisels – they are ground at 25°, and sharpened on an oilstone at 30°. Note, however, that turning chisels are sharpened from both faces whereas bench chisels are sharpened from one face only. The angle of sharpening for scraper chisels is between 60° and 70° to the upper faces of the tool. All turning tool handles are made from ash, hickory or beech. The metal ring at the blade end of the handle – the *ferrule* – is made of brass and the blades are made of good quality carbon tool steel.

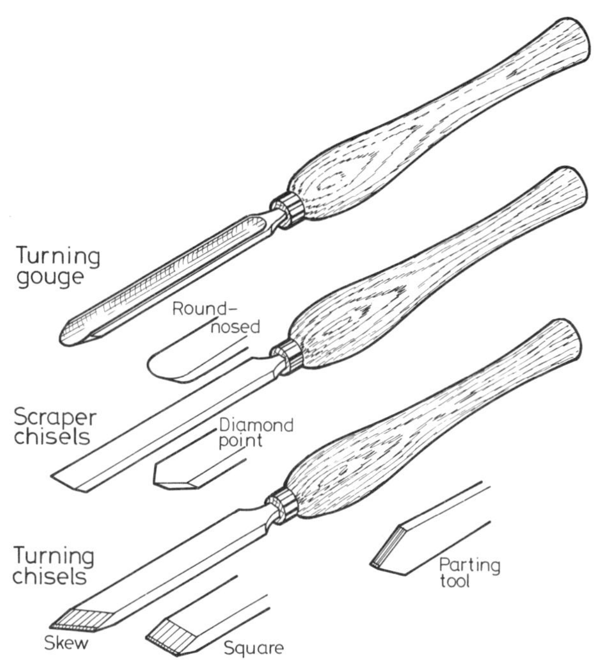

Hand turning-tools for wood

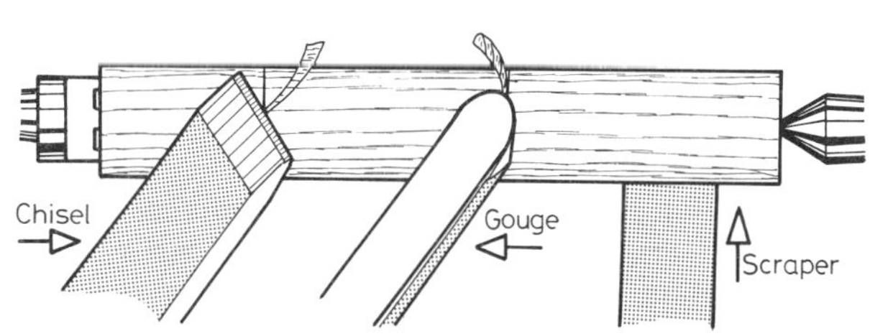

Methods of using hand turning-tools for wood

Hammers and mallets

Hammers

Hammer heads are generally made of cast steel with ash or hickory handles – the wooden handles pass through *eyes* in the hammer heads, to be held firmly in position with steel and wooden wedges. Some modern hammers have steel and fibre glass handles, which are held to the heads with epoxy resin. Hammer handles are of a shape enabling them to be held firmly and of a length suitable for delivering efficient blows. When using any hammer make full use of the handle length by holding the handle at its end and not holding it near to the head.

The working ends of hammer heads are known as *peins*, and these should be kept clean in a bright and polished state, free of dirt, grease and spots of hardened glue. In particular, peins of hammers used for planishing must be brightly polished to avoid blemishing the metal surfaces on which they are used. When a hammer handle becomes damaged or when it becomes loose in the head, replace it or re-wedge it without delay. Serious injury can occur if a head flies from a handle because the head is not properly secured.

Hammers for woodworking

Three types of hammer are commonly used when woodworking; the *Warrington* pattern carpenter's hammer; the *pin hammer* and the *claw hammer*. The most frequently used type is the carpenter's hammer which is made with head weights of from 170 g to 450 g, although a 350 g head is the most suitable for general work. Pin hammers are of the same pattern, but with narrower heads weighing only 100 g. Both carpenter's and pin hammers have a flat round pein, for driving nails in place, and a cross pein for starting off thin nails and pins held between the fingers of one hand. Pin hammers are made for driving light nails and pins in place. The third common hammer for woodworking is the claw hammer, often with a 620 g head. This heavier hammer is for heavy nailing, the claw being suitable for removing nails which have become bent as they are driven home.

Hammers for metalworking

Many different types of hammers are made for working in metals. Only a few of this large number are illustrated here.

The general-purpose metalworking hammer is the engineer's hammer with ball and round peins and heads weighing between 200 g and 900 g, a common bench size being 350 g. Lighter and heavier engineer's hammers, however, are necessary when working in metals. Engineer's hammers are for use with dot and centre punches, with cold chisels for riveting and for shaping metal while it is cold. They are also the general-purpose hammers for forging hot metal to shape. A smaller 100 g head hammer with a pin ball pein is made for light work.

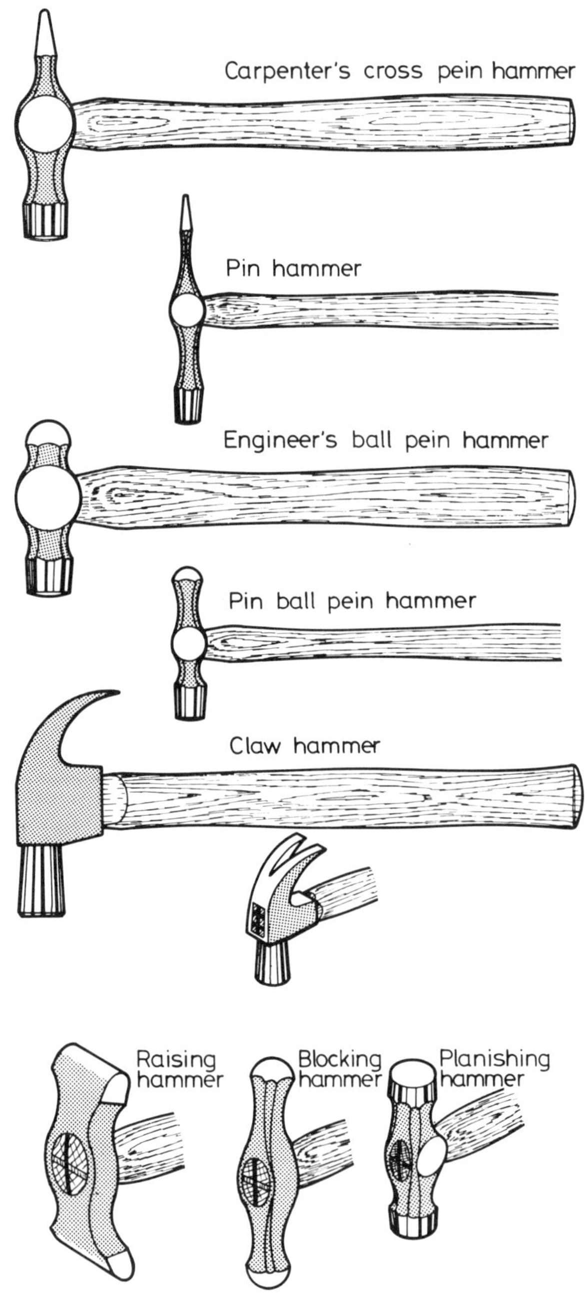

Woodwork and metalwork hammers

The other hammers illustrated here are for raising or hollowing sheet copper, sheet gilding metal and sheet silver. The three main groups of hammers for this work are *raising, blocking* and *planishing* hammers. Raising and blocking hammers are made with relatively large heads of about 350 g, but planishing hammers, made in a large number of different pein shapes, are usually made with heads of about 120 g.

Many other types of hammer are seen in metalwork shops, for example, *sledge hammers* with heads of from 2 to 6 kg and, at the opposite extreme, *repoussé hammers* with heads of about 80 g. Sledge hammers are for heavy forging work and repoussé hammers are for use with repoussé punches in decorative sheet metalworking.

Mallets

As with hammers, a larger range of mallets is necessary when working in metal than when working in wood. Mallets are commonly made with heads of woods such as beech or box with ash, hickory or cane handles. Apart from carpenter's mallets, in which the handle is of tapered shape fitting into a tapered hole in the mallet head, handles are commonly fixed in place by wedging.

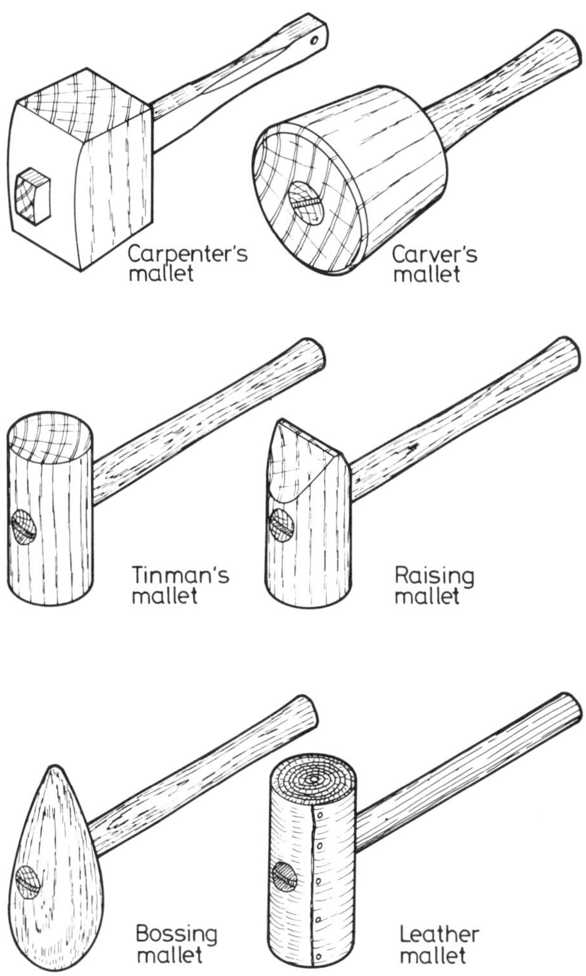

Woodwork and metalwork mallets

Mallets for wood

Only two mallets are commonly found in woodwork practice. These are the *carpenter's mallet* for general bench work and the *carver's mallet* for use with carving tools. The beech head of a carpenter's mallet is either 115 mm or 130 mm long and that of the carver's mallet is either 115 mm or 130 mm in diameter. The handles are of ash or hickory.

Mallets for metalwork

A number of different types of mallet are made for working metal, three of which – *tinman's, raising* and *bossing* – are shown in drawings. These are made with heads of boxwood into which handles of ash, hickory or cane are wedged. The names of these three mallets define their uses.

The fourth mallet illustrated here is a *rawhide mallet* with a leather head and a cane handle. The soft head of rawhide mallets is less liable to cause surface marking. Another soft-faced mallet, not illustrated here, is one in which replaceable faces of nylon, copper or rubber can be fitted.

Examples of hammers and mallets in working conditions

1. A claw hammer for pulling a bent nail from a frame.
2. A tinman's mallet for bending sheet metal.
3. A bossing mallet forming the shape of a sheet metal bowl.
4. Raising a vase-shape from sheet metal with a raising hammer.

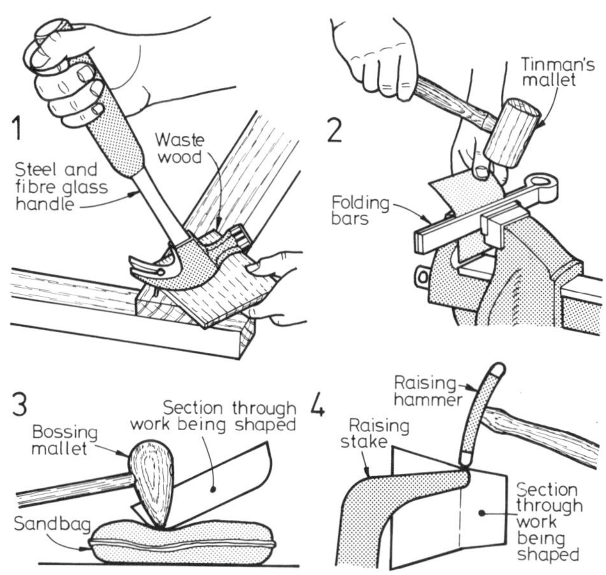

Hammers and mallets in use. 1. Pulling out a nail. 2. Bending sheet metal. 3. Sinking a copper dish. 4. Raising a vase.

Pliers spanners, and grips

Many members of this large group of tools have specific and limited applications. Only those tools most commonly used are illustrated here.

Pliers

Combination pliers

These tools are the most common type of plier as they are very versatile. Combination pliers can not only hold flat or cylindrical items firmly in the serrated jaws but the cutters in each jaw can cut soft metals up to 2 mm in diameter. For electrical work the handles may be insulated with sleeves, which also make holding and gripping easier.

Round and flat-nosed pliers

Round-nosed pliers have either half-round jaws serrated along the inner surface, (used to grip small items), or completely round noses to form circular loops for example, in fitting wires to electrical terminals. They are made with different nose lengths.

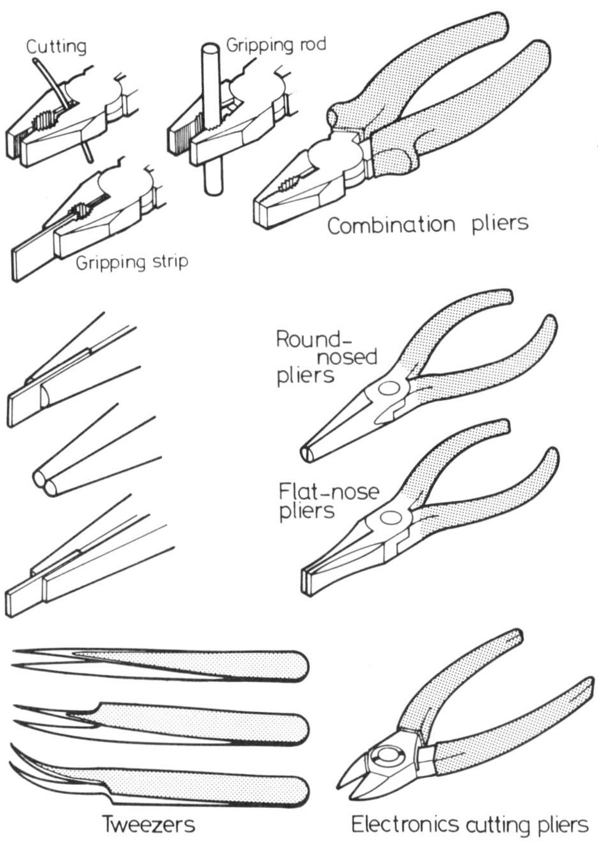

Types of pliers

Electronics cutting pliers

A range of pliers is manufactured for electrical and electronics circuit work. The cutting pliers illustrated allow the accurate cutting of wires to length and they are fitted with insulating sleeves as a protection against shocks from the circuitry being built.

Tweezers

Three styles of precision tweezers are shown. They are used for precise work, such as placing small electronics components in a circuit or assembling small machines.

Spanners and grips

Spanners

These tools are designed to tighten and loosen nuts and bolts. They are made in a wide range of sizes and types, of which only two are shown here. Other common types are ***adjustable spanners, socket spanners*** and ***box spanners***. The size is indicated at the ends of every spanner. Common groups of spanners are AF (across flats), metric and Whitworth.

Mole grips

These have a lever mechanism which is adjusted to exert and maintain a vice-like grip on the item being held. This leaves the hands free for other work. Various sizes and types of jaw are available.

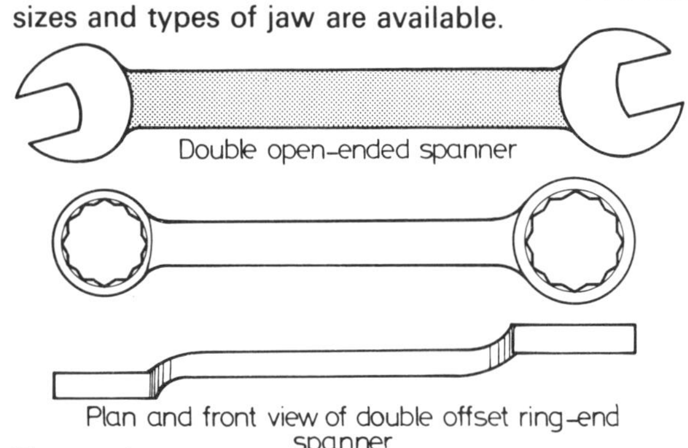

Types of spanners

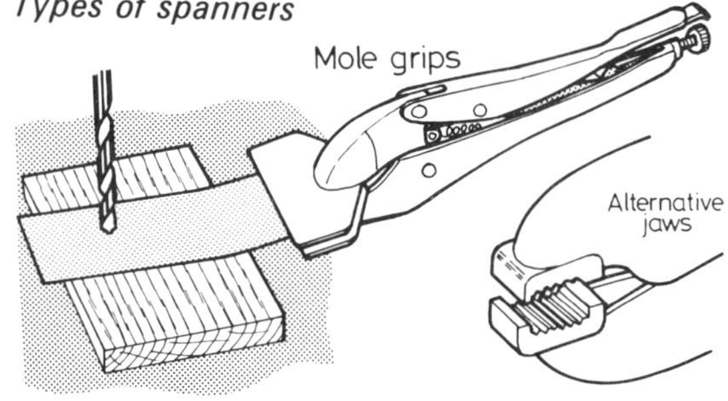

Mole grips

Drills, bits and boring

Twist drills

Twist drills are for boring holes in woods, metals or plastics and they are generally made in the two styles illustrated. Those with straight parallel shanks are designed to be held in 3-jaw drill chucks of hand or machine drills. They are mainly for drilling holes up to about 13 mm diameter and a good range of straight-shank drills would be from 1 mm to 13 mm diameter, in steps of 0.5 mm.

Morse taper-shank drills are designed to be fitted into the Morse taper sockets of lathe headstock spindles, toolstock spindles or drilling machine spindles. Holes of up to 30 mm diameter can be bored with Morse taper shank drills.

Twist drills are really designed for boring holes in metals. When drilling in wood, they are suitable only for holes up to about 6 mm diameter as larger holes are best bored with purpose-made *wood bits* (see below). Twist drills in the smaller diameters are also very suitable for boring holes in plastics such as nylon, acrylics, polypropylene. When drilling by machine in any material, adjust the speed of the machine to one suitable for the drill size and the material. In general, the smaller the hole, the faster the speed; the harder the material the slower the speed.

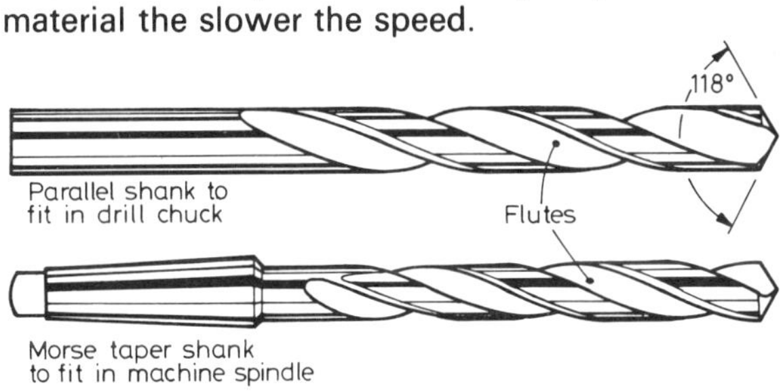

Details of twist drills

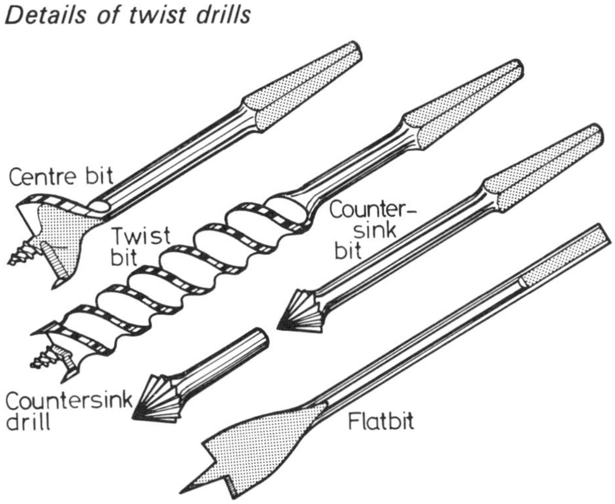

A variety of bits for boring holes in wood

Wood bits

Bits are for boring holes in wood. Holes up to about 30 mm diameter can be bored with *centre bits, twist bits* or *flatbits*. Bits with square taper shanks are for use with a carpenter's brace and those with straight, cylindrical shanks are for use in a hand drill. The flatbit range of bits is for use in machine-powered drills. Countersink bits and drills are for countersinking holes to receive the head of countersunk screws.

Tools for boring large diameter holes

Tank and washer cutter

With its square taper shank fitted in the chuck of a carpenter's brace, this tool can be adjusted to cut holes from 25 mm diameter up to 125 mm diameter in a variety of sheet materials such as metals, plastics, leather, rubber. Its two cutting blades can also be adjusted to cut washers up to the same diameter.

Hole saw

This is designed for fitting into the chuck of an electric drill and it will saw out holes from 20 mm to 70 mm in diameter with its interchangeable saw cutters, which will bore holes in wood equally as well as in other materials. The waste circles from this tool, when it has completed boring holes in wood, can be used as wheels in, say, a child's wheeled toy.

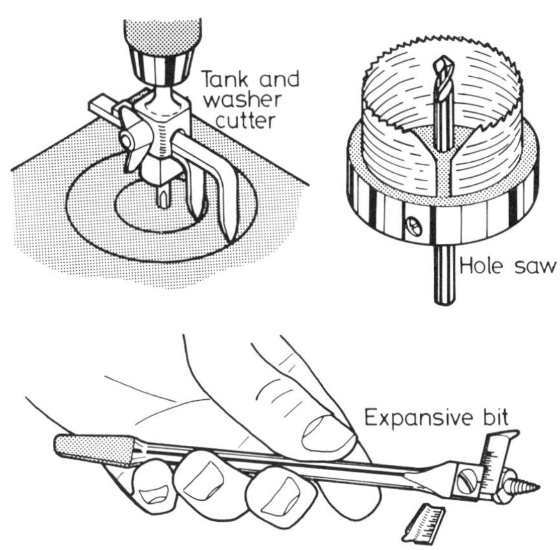

Tools for boring large diameter holes

Expansive bit

Another bit for fitting into a carpenter's brace for boring large diameter holes in wood is the *expansive bit.* Two sizes of this bit are made, each with two cutters: the smaller will bore holes from 13 mm up to 30 mm in diameter; the second is for holes between 22 mm and 75 mm diameter. The cutters are adjustable along a dovetail-like groove system in which they can be locked at any required position using a locking screw.

Tools for boring small diameter holes in wood

Bradawls

Two types of *bradawl* are illustrated. These are both tools used with hand pressure to bore small holes in wood, such as the holes required when screwing or nailing parts together. Standard bradawls are made with chisel-like edges which, when rotated first one way and then the other, will cut small holes. The second type of bradawl is made with a square tapered blade finishing with a sharp point. As this tool is rotated about a point in wood, under hand pressure, the sharp square edges of the blade cut away a hole. Care is needed when boring holes with bradawls at the end of wood because the short grain will split open if the hole is bored too quickly or with too great a pressure placed on the handle. Bradawls can be purchased in a range of sizes suitable for boring holes up to about 3 mm diameter.

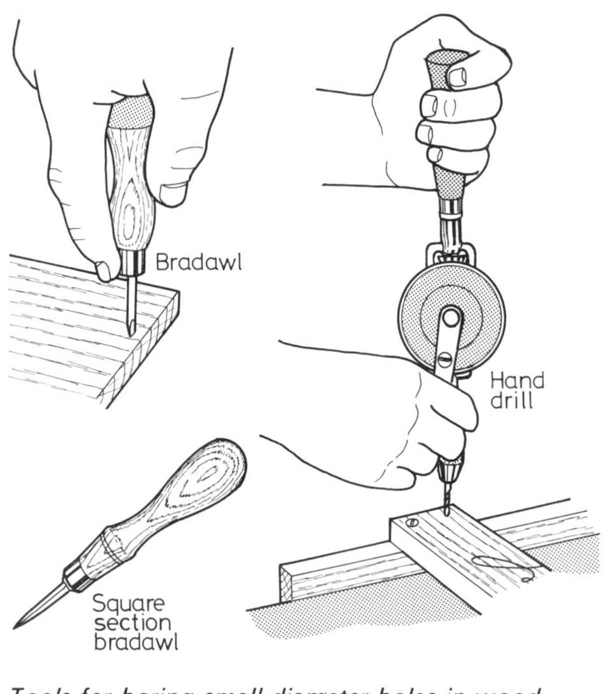

Tools for boring small diameter holes in wood

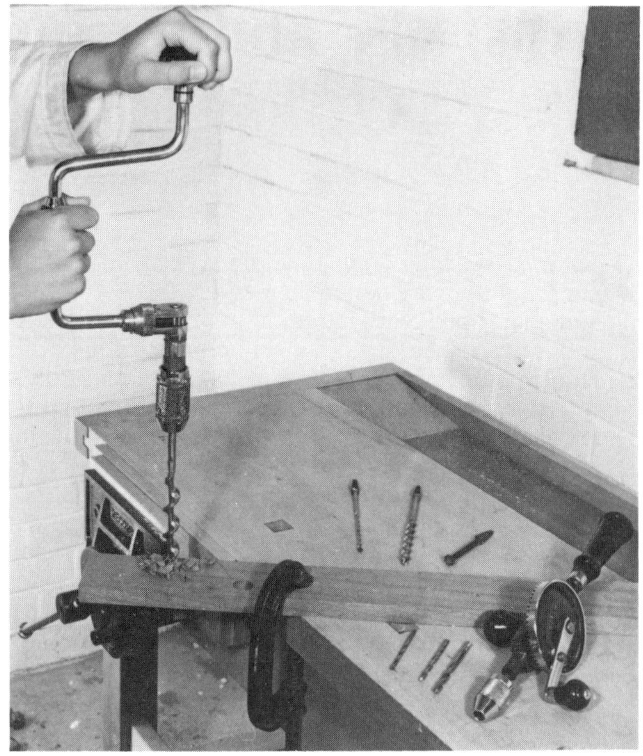

Boring a hole in wood with carpenter's brace and bit

Centre drill

Centre drills, sometimes referred to as *Slocombe* drills, are for boring holes in the centres of the ends of rods which are to be turned on a lathe by being held between the lathe centres. Further reference will be made to centre drills on page 55. Centre drills are a combination of a small and short drill which bores a pilot hole, and a countersinking drill which bores a countersink at an angle of 60° in the end of the pilot hole. The 60° countersunk holes will fit snugly over the points of the lathe centres and the small pilot hole prevents the lathe centre from having to rest directly on to the end of the countersunk hole, thus reducing friction as the work rotates around the lathe centre.

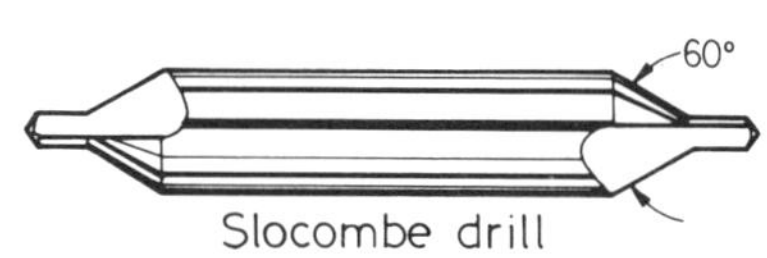

A centre or Slocombe drill

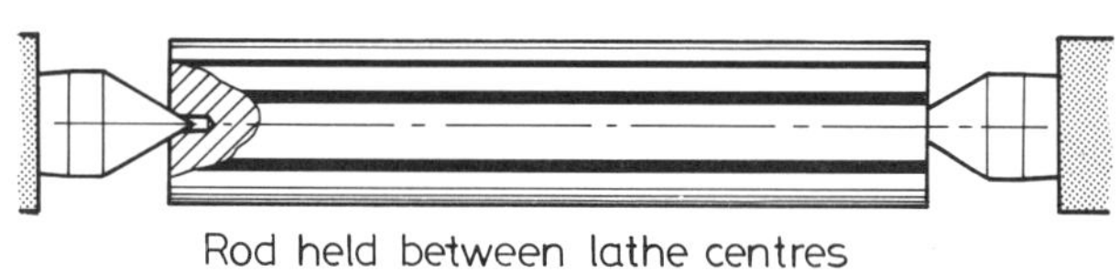

Holes bored in end of rod with centre drill

Exercises

1. A number of different types of vice are made. Name three types, state from what materials they are made and specify uses for which they are made.
2. Make sketches of: a funnel stake; a planishing stake; a *horse* in which planishing heads can be fitted. What are these stakes used for?
3. A number of different jigs are used for holding wood when it is cut to length, to ensure that its ends are either square or at 45° to the length. Name three such jigs and state the exact purpose for which each is made.
4. Why are vice clamps or vice jaws used with engineer's vices? From what materials are they made and why are wooden vice jaws fitted to most vices of woodwork benches?
5. What are toolmaker's clamps? Give three tasks for which they are suitable.
6. What are the differences between try squares made for wood and those made for metal? Which marking tools are commonly employed when using these try squares?
7. What are odd-leg dividers? What are they used for? What are dividers? What are dividers used for?
8. Three types of gauge are in common use for marking lines on wood. Name them, and make sketches of all three gauges. What is each made for?
9. Explain, with the aid of sketches, the differences between a centre punch and a dot punch. Give uses for each punch.
10. What are face marks? Why is it important to make face marks on materials being worked? Give examples of work involving face marks.
11. If you were sawing shapes from acrylic sheet of 3 mm thickness, what precautions would you observe and why?
12. Sketch the following saw teeth:
(a) rip; (b) crosscut.
What is the purpose of making some saws with rip teeth and others with crosscut teeth?
13. Explain, with the aid of sketches, the different methods of tensioning the blades of:
(a) a hacksaw; (b) a coping saw; (c) an Abrofile.
14. Sketch a piercing saw and explain the purpose for which this saw is made.
15. Explain, with examples, what is meant by *sawing into the waste*. Why is it important to learn how to do this?
16. At what angles are the cutters held in the bodies of the planes:
(a) jack; (b) shoulder; (c) block; (d) smoothing; (e) router.
17. What is meant by the terms *fence* and *depth stop* in connection with planes?
18. Name four *cuts* of files and give examples of the uses suitable for each of the cuts you have named.
19. State three safety rules applicable to using files when filing materials. Explain the reasons for the rules you have stated.
20. What is the difference between an engineer's file and a needle file? Give examples of processes for which you would use (a) an engineer's file, (b) a needle file.
21. What is meant by *draw-filing*? Give examples of work which should be draw-filed.
22. Several different types of tool can be used when you wish to cut a curve in wood. Name three types of tool for the purpose. Give examples where you would use each of the three types of edge-shaping tool.
23. What are the differences between the following chisels:
(a) a square-edged firmer chisel;
(b) a bevel-edged firmer chisel;
(c) a mortise chisel.
Stae reasons for the differences.
24. What material are wood chisels made from? Why are the materials you have named used for making chisels?
25. Three types of cold chisel can usually be found in the school workshop. Make a sketch of each. For what purposes are each of these chisels used?
26. What is a gouge? Name three types of gouge. Give uses for the gouges you have named.
27. What are the differences between the following hammers:
(a) carpenter's; (b) pin; (c) ball pein; (d) raising.
From which materials are hammers made? Give reasons why these materials are used.
28. How are hammer heads held to their handles? Use sketches to elaborate your answer.
29. The following materials are used for making mallets:
(a) box; (b) ash; (c) rawhide leather; (d) beech.
Which type of mallet is made from each of these materials?
30. Sketch and name uses for each of the following tools:
(a) carver's mallet; (b) raising mallet; (c) bossing mallet.
31. What is the general cutting angle of a twist drill?
32. Sketch each of the following and state a purpose for each:
(a) twist bit; (b) centre bit; (c) countersunk bit; (d) flatbit.
33. Why is it necessary to bore holes in the ends of pieces of metal which are to be turned 'between centres' in a lathe?

Screwdrivers, screws, nuts, bolts and rivets

Screws, nuts, bolts and rivets are among the most common and effective items for securing the parts of constructions to each other.

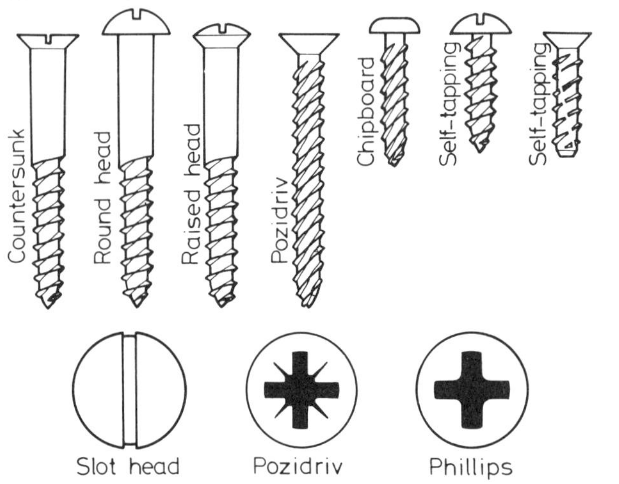

Wood screws

Wood screws

Most wood screws are made of bright mild steel although brass, aluminium and copper screws are made. Wood screws may also be purchased with anti-rust coatings of zinc, chromium and cadmium, as well as with decorative coatings of brass, bronze and black lacquer. Wood screws are measured by their effective length (parts that are within the wood), and by their gauge diameter. Common gauge sizes are 00 to 20 (about 1 mm to 7 mm) although larger screws up to 50 gauge are made and lengths range from 5 mm to 50 mm. Details of the length, gauge, head shape, metal and coatings are necessary when buying screws.

90% of all wood screws used are *steel-countersunk* for general-purpose screwing. *Round-head* and *raised-head screws* are for screwing thin metal fittings to wood. *Pozidriv screws* (usually zinc or cadmium coated) have twin-start threads for rapid insertion; they are used for general household joinery, such as window fittings. To screw into chipboard, *twin-start chipboard screws* provide a stronger hold than wood screws. *Self-tapping screws*, made of hardened steel, are for screwing into thin sheet or soft metal; they may have tapping slots cut along the thread. The slot on the screw head may be straight, *Pozidriv* or *Phillips*, as shown, so the correct screwdriver is necessary.

Screws and bolts

When referring to machined screws such as those shown, common practice is to call those threaded throughout their length *screws* and those threaded part-way along their length *bolts*. Machine screws and bolts, together with their nuts and appropriate washers, are mainly for holding metal parts to each other. This does not exclude their being used for bolting wooden or plastics parts to each other where a design warrants such methods. The common screw thread for such bolts is the Metric thread, but screws and bolts with Whitworth, British Association and British Standard fine threads will be seen.

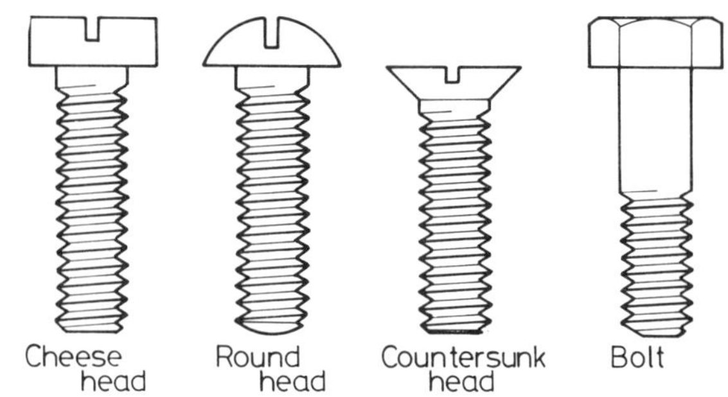

Screws and bolts

Rivets

Rivets are commonly made from mild steel but copper and aluminium rivets are available. The three head shapes most likely to be seen are those shown in the drawings here. *Bifurcated rivets* are for riveting leather and rubber and they are made from copper, or from copper-coated mild steel. Bifurcated rivets may need washers under the bifurcated part when it is spread out to make a joint. 'Pop' rivets are made by the firm Tucker Fasteners Limited and are a proprietory form of rivet which can only be fixed in position with the aid of a 'Pop' riveting tool. 'Pop' rivets are valuable for securely fixing parts to each other where one side of the join is inaccessible.

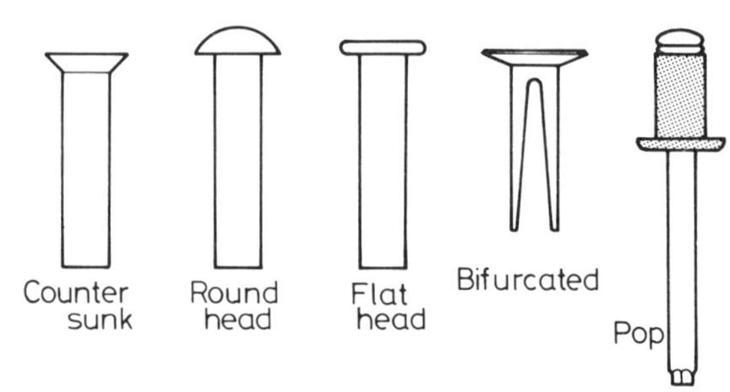

Rivets

Screwdrivers

There are sizes of screwdrivers suitable for every type of screw. Screwdrivers with plastic handles never become loose from the blade, and they are made in a variety of different shapes. The blade tips of screwdrivers are made in shapes and sizes to fit the different screw slots. Pozidriv and Phillips screwdrivers have graded tip sizes from 1 (the smallest) to 3. It is important to select the proper tip size for any screw to avoid damaging the slot.

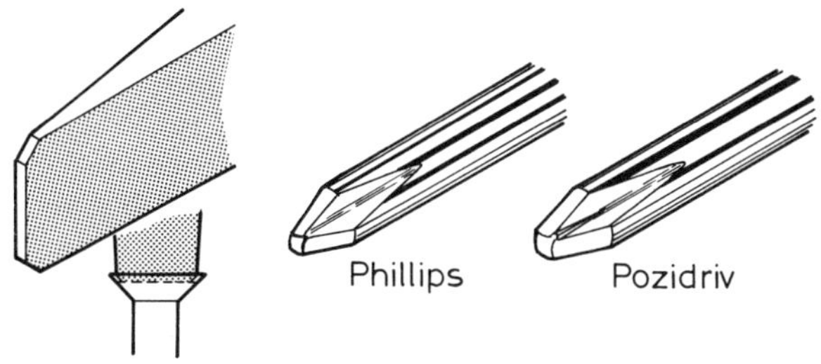

Screwdriver tips

Making a screwed joint in wood

First select a suitable length and gauge of screw. The length should be such that about $\frac{2}{3}$ of the screw is in the lower part of the join. The gauge diameter will depend upon the strength expected of the join. Five drawings and the text below describe the stages in making a good and secure screwed joint.

1. Drill holes in the upper piece with a drill of shank diameter as follows: for gauge 10, 6 mm; for gauge 8, 5 mm; for gauge 6, 4 mm; for gauge 4, 3 mm.
2. Countersink the holes to receive the screw heads.
3. Position the upper piece on to the lower piece and, with a pilot drill or a bradawl of the same diameter as the solid part of the screw thread, bore pilot holes to the full depth of the screw: for gauge 10, 4 mm; for gauge 8, 3 mm; for gauge 6, 2 mm; for gauge 4, 2 mm.
4. This shows a section through the shank hole, its countersink and the pilot hole.
5. This is a section showing the screw driven fully home.

If the procedure above is followed, screwed joins can be made to give maximum strength without any danger of the parts holding the screws being split or otherwise damaged.

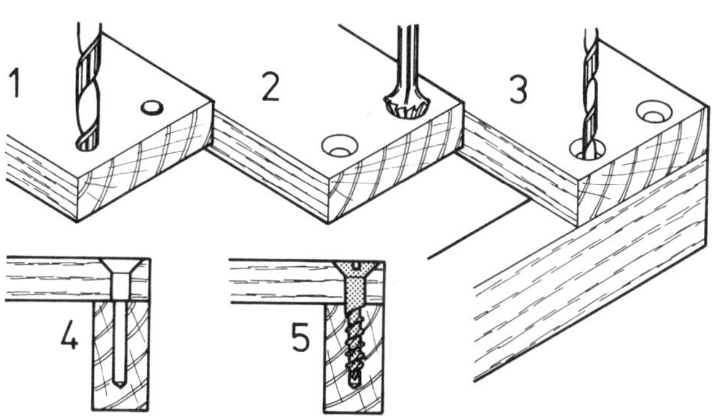

Stages in screwing two pieces of wood together

Some methods using machine screws and bolts

The drawings show sections through five different assemblies involving machine screws or bolts and their associated washers and nuts. Note that in all examples, washers are placed under that part of the nut or bolt which must be rotated in order to tighten the parts together.

1. Two sheets of metal are bolted together with a cylindrical spacer between.
2. Two pieces are bolted together with a machine bolt. The hole in the lower part has been tapped to receive the screwed part of the bolt (see page 34 overleaf).
3. This shows a spindle fitted into a hole in which it is to rotate. In order to prevent the spindle nut working loose as it rotates, a lock nut (the thinner of the two nuts) locks the two nuts tightly against each other. Two washers, one each side of the hole, allow free spindle rotation.
4. Two parts assembled with a machine screw and a *wing nut* are shown. Wing nuts allow speedy dismantling without the need to find and to use a spanner.
5. This is a section through a spindle held in a ball-bearing in which it can freely rotate. A wheel is bolted on to the spindle by means of a *castle nut*. The castle nut is prevented from revolving on the spindle with a *split pin* which is pressed through one of the slots in the castle nut and through a hole in the spindle.

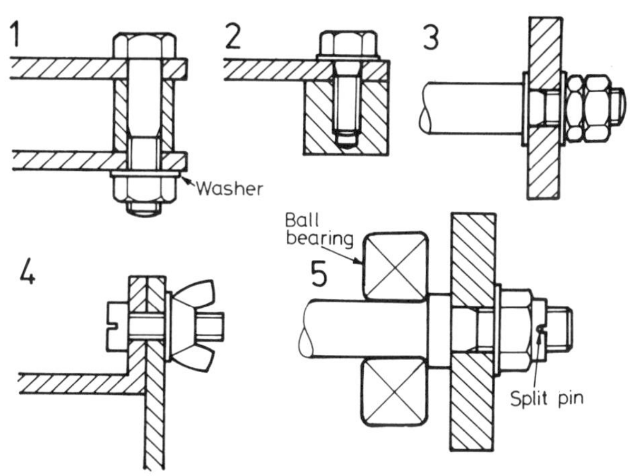

A variety of methods of using nuts, bolts and washers

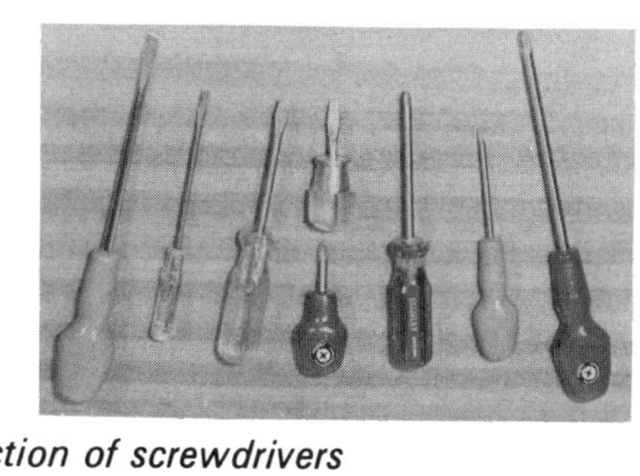

A selection of screwdrivers

Cutting screw threads in metals and plastics

Screw threads up to about 12 mm diameter can be formed with taps (for internal threads) and dies (external threads) held in tap wrenches and die stocks. ISO metric M and other thread forms can be cut. A set of ISO metric taps and dies would cut threads of sizes M3, M4, M6, M8, M10, M12.

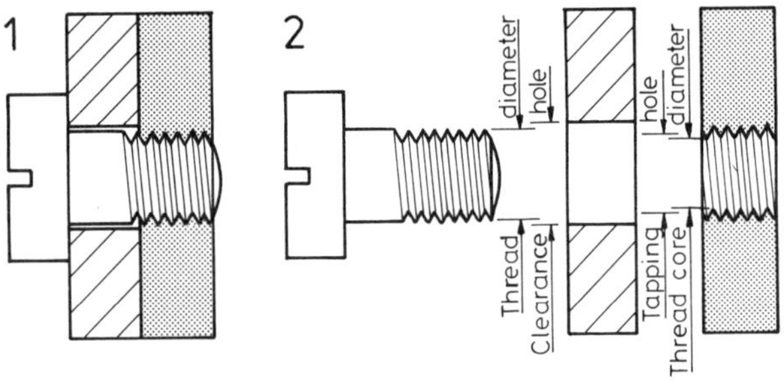

Terms connected with screwed holes

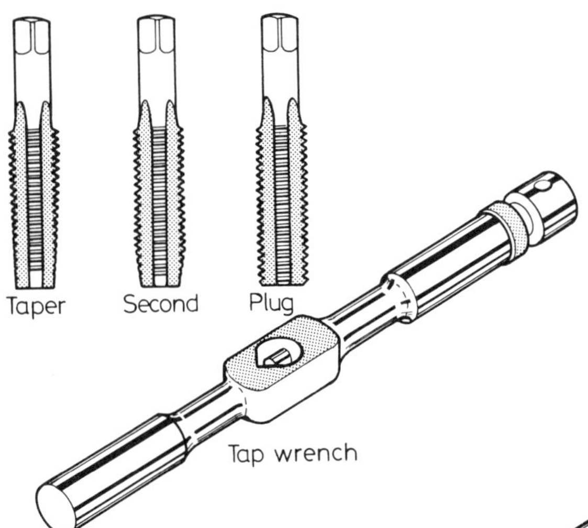

Taps and tap wrench

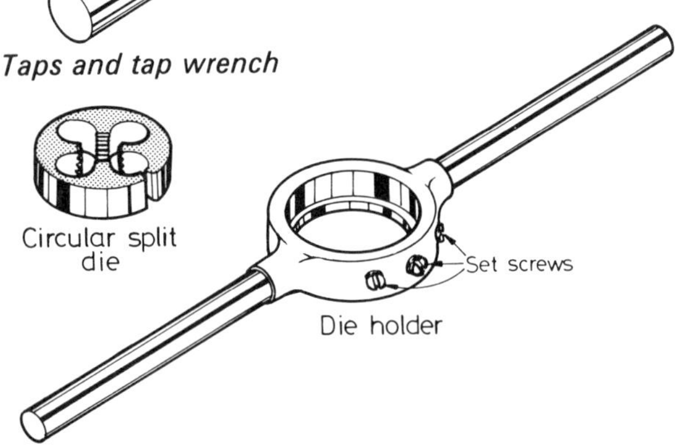

Die and die holder (or stock)

Tapping a hole

Forming screw threads in a hole is called tapping. A taper tap, second tap and plug tap of the same thread size make up a set of taps. The taper tap, shaped to start cutting a thread, is used first and followed, for a blind hole, by the second tap and then the plug tap. If used carefully, the plug tap can follow the taper tap. To tap right through a hole, the thread can be completed by the second tap.

Before tapping, a hole must be bored with a drill of tapping hole diameter. Drill sizes for a range of ISO metric screw threads are as follows:
M3, 2.5 mm; M4, 3.3 mm; M6, 5 mm; M8, 6.8 mm; M10, 8.5 mm; M12, 10.2 mm.

A set of tapping drills should be available to ensure that correct sizes of tapping holes are bored.

Stages in tapping a hole

1. Bore a hole with the correct size of tapping drill.
2. Check that the tap is in line with the hole. Using a suitable lubricant, rotate the tool clockwise and anticlockwise until the whole length of thread required is cut.
3. If tapping right through a hole, the thread can usually be completed with a second tap. If tapping a blind hole, the second tap must be followed with a plug tap.

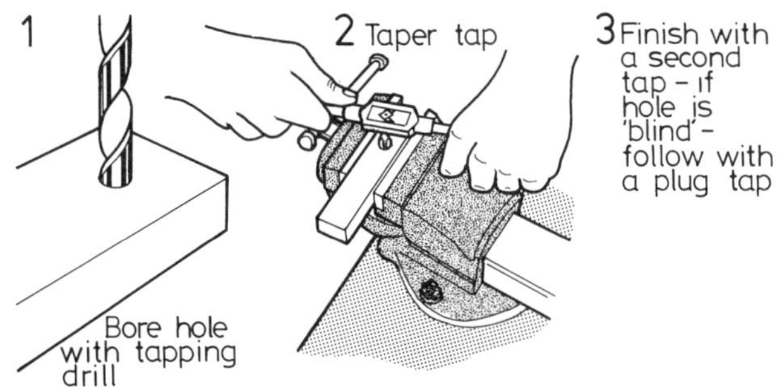

Stages in tapping a hole

Precautions when cutting screw threads

1. Taps are hard and can break if used with unnecessary force, or without proper caution. To remove a broken tap from a hole can be difficult, quite apart from a waste of an expensive tool.
2. Dies can be adjusted with the aid of the three set screws in the die holder or stock. Taps are not adjustable. Because of this, always tap an internal thread before cutting its external thread with a die. The die can then be adjusted to cut a thread which fits the tapped thread with a good degree of accuracy.
3. Both taps and dies should be worked with a forwards and backwards rotation. First rotate the tool clockwise to cut perhaps half the circumference of a thread. Then rotate the tool back anticlockwise, more than half a circumference, to remove the waste cut by the tap or die. Repeat this clockwise and anticlockwise rotation throughout the whole thread cutting action.
4. A lubricant will be required when cutting threads in some metals. For example: steels – thread cutting grease; aluminium and copper – paraffin or a light mineral oil; brass and plastics – no lubricant.
5. Before starting to cut a thread make sure that the tap or die is at right angles to the hole or rod on which the thread is to be cut.

Stages in cutting an external thread

1. File a slight taper on the end of the rod on which the thread is to be cut. Place the correct size of die in its holder and tighten the central screw of the three set screws. Then tighten the other two set screws – this will ensure that the die will cut as large a diameter thread as the die is capable.
2. Place the stock and die in position on the rod end the right way up, with the die secure under its flange in the holder. Check that the die is square with the rod from both front and side.
3. Use a lubricant, and rotate the tool in a clockwise direction, then an anti-clockwise direction, cutting about half of one thread each time. Repeat until the whole of the required thread has been cut.
4. Check if the cut thread fits into its tapped hole. If it is too tight, then loosen the central set screw of the holder, and tighten the other two screws. This makes the die ready for cutting a thread of slightly smaller diameter. Re-cut the thread, and repeat this process until the external thread fits as required in the tapped hole.

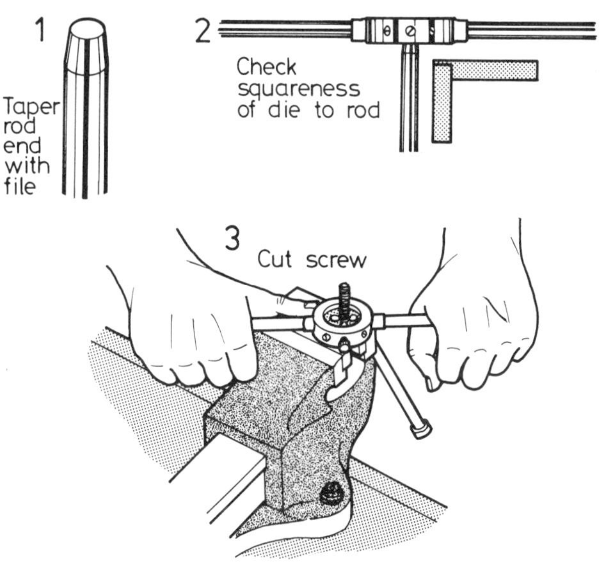

Stages in cutting a screw-thread

Riveting

'Pop' rivets

One of the easiest methods of riveting, yet one which makes a very effective joint, is to use 'Pop' rivets which must be secured in place with the aid of a 'Pop' riveting tool. Select the correct size and type of 'Pop' rivet and load it into the riveting tool. Then bore holes of the required rivet size in the parts to be riveted and place the parts together. Position the tool with its rivet over and through a hole and close the tool's handles so that the rivet is securely fastened.

'Pop' riveting is suitable for joining sheet metals and, if the correct rivets are used, for riveting sheet plastics.

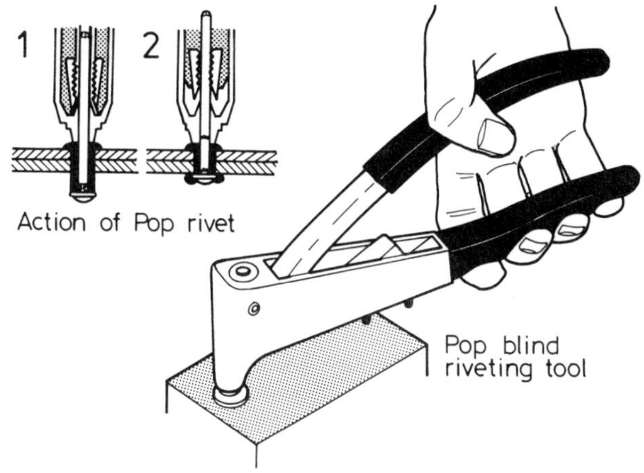

'Pop' riveting

Hammer riveting

When thick metal sheets are to be joined together, rivets which are hammered in place are suitable. The three common types of rivet illustrated on page 32 are the *countersunk-head rivet*, the *round-head rivet* and the *flat-head rivet*. Flat-head rivets are for joining sheet metal which is too thin to be countersunk. A series of sectional drawings, numbered 1 to 5 illustrate the stages, described below, in fixing a round-head rivet joint.

1. Bore holes of the same diameter as the rivet in the parts to be riveted and push the rivets into place through the holes. Clamp a rivet snap in a bench vice and place the half-round head of the rivet in its socket. With the aid of a rivet set and a hammer, force the two sheets together over the rivet by hammering down on the head of the rivet set.
2. Saw off the end of the rivet to leave a length of about $1\frac{1}{2}$ times the rivet diameter standing above the surface.
3. With the ball pein of an engineer's hammer, hammer the protruding part of the rivet to an approximately half-round shape.
4. This shows the hammered rivet prior to shaping with a rivet snap.
5. Shape the hammered rivet to a good half-round shape with a second rivet snap.

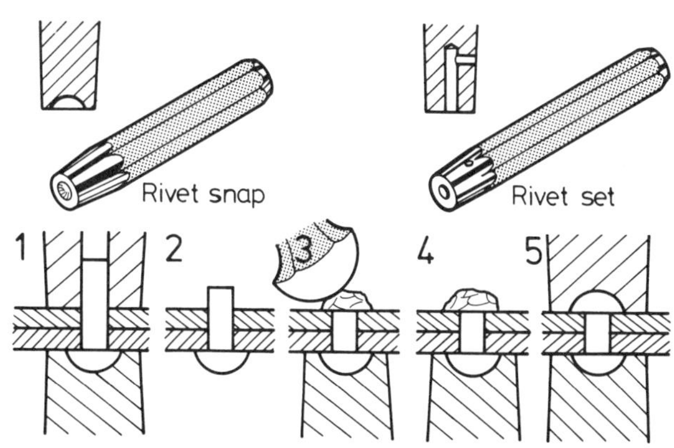

Stages in riveting

Cold and hot bending

Bending thermoplastic sheet

Thermoplastic sheet material can be uniformly bent along a straight line and through a small radius if a narrow strip of the sheet along the bend is heated to a temperature appropriate to the plastic being worked and strip heaters are made for this purpose. A strip heater manufactured by C. R. Clarke and Company of Ammansford, Dyfed, is illustrated here. Thermoplastic sheet placed on the support bars of this heater is quickly heated in a concentrated line to a temperature which will allow uniform and controlled bends to be rapidly produced. Bends of larger radii can be produced by increasing the gap between heater and the work by adjusting the positions of the support bars. Simple jigs for holding the thermoplastic sheet in its bent position while it cools are also illustrated. These may consist merely of nails in a piece of wood into which the bent part is placed, or they may include purpose-made wooden formers to which the sheet is clamped and removed when it becomes cold.

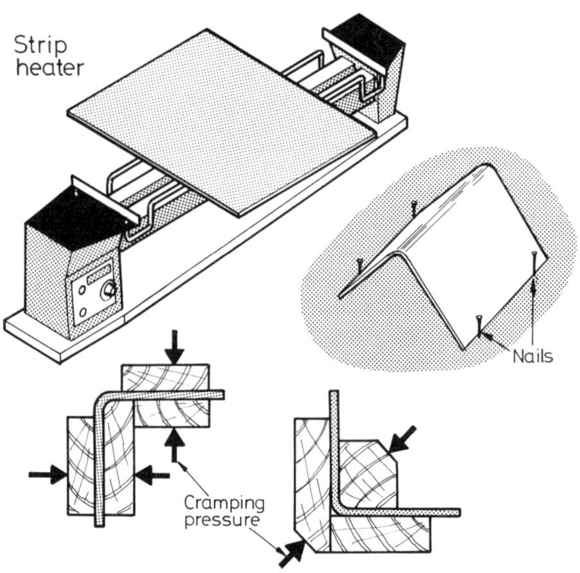

Bending thermoplastic sheet

Vacuum forming

This is a method of forming three-dimensional shapes from thin thermoplastic sheet. The sheet is firmly clamped flat between the sides of some form of holder and then warmed by heaters until it is in a rubber-like, malleable state. The heaters are swung to one side and the warmed, malleable sheet is brought into contact with a mould by forming a vacuum between the sheet and mould. The vacuum is produced from a vacuum pump, either a mechanical or a hand pump, and the resulting pressure forces the sheet to drape exactly over, or on to, the mould. Air cooling takes place rapidly and when the article is cool it can be taken from the mould. *Male* or *female* moulds (see page 48) are used, depending on which is the easier to make.

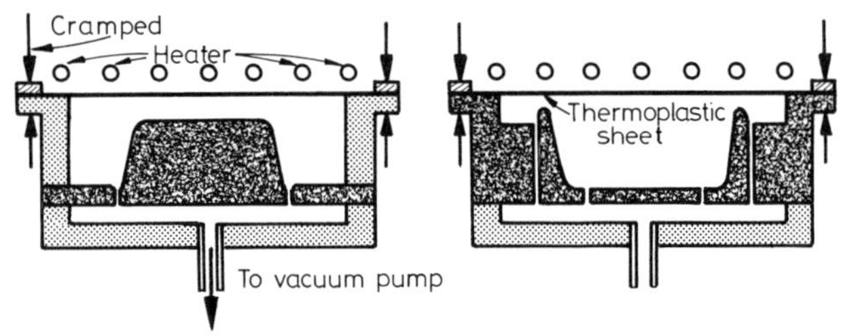

Diagrammatic drawings showing vacuum forming of thermoplastic sheet

A vacuum forming machine for thermoplastic sheet

Parts of a school-made vacuum forming machine are illustrated (page 37). This machine consists of a wooden box topped with pegboard and is connected to an electric vacuum pump via plastic tubing. The heater, which was made from lengths of Eureka resistance wire connected to a 12 volt power supply, is not shown in the illustrations. Moulds are placed on the pegboard and when the thermoplastic sheet has been heated to a malleable state, the electric pump is switched on forcing the now malleable sheet held in the angle-iron frame on to the mould's surfaces. This vacuum forming machine is capable of forming mouldings from sheet of up to 1.5 mm thick – pvc, acrylic and polypropylene sheet being suitable. A small industrial vacuum-forming machine is shown in a photograph on page 36, opposite. This machine was developed for the shaping of parts of footwear by the firm Satro of Kettering.

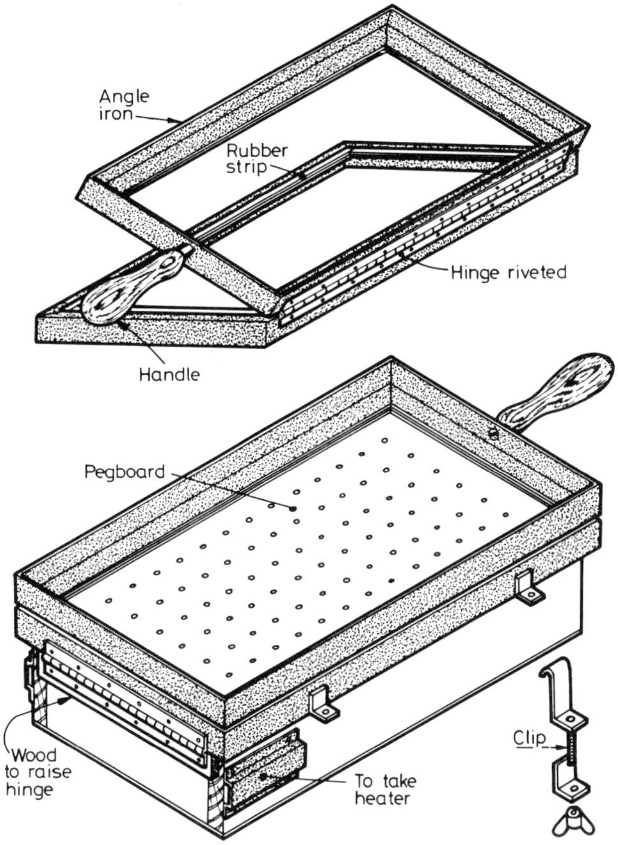

A school-made vacuum forming machine

Beaten sheet metalwork

Sheet metals such as copper, gilding metal and silver can be beaten to shape on stakes with hammers and mallets. Three examples of beaten metalwork are shown in the photograph here – two vases and a small tray. The bodies of these three items were either raised or sunk from circular sheets of gilding metal; then feet were soldered to the vases, and a handle soldered to the tray.

Tools for beaten metalwork

Some of the tools employed for beaten metalwork have been described on page 8 (stakes) and pages 26 and 27 (hammers and mallets). One of the problems involved in this form of craftwork is that a large number of different shapes of stakes, stake holders, hammers and mallets are needed. Only a few of this large number of tools can be shown in this book. However, a sufficient number are illustrated to show the principles of work involved in beaten metalworking.

Tin-snips or shears

Tin-snips are for cutting the shapes required from sheet metal. Despite the common name of tin-snips, these cutters will deal adequately with sheet copper, brass, gilding metal and silver up to 16 Standard Wire Gauge thickness (about 1.5 mm). Snips are made in a number of lengths from 200 mm up to 350 mm overall, and they are also made with curved blades for cutting along curved lines.

Stakes

The sheet metal being worked is placed on stakes while being beaten with mallets and hammers. The working surfaces of stakes are polished to a fine and smooth finish to avoid causing blemishes on the surfaces of the metal as it is hammered into shape. The bright, smooth surfaces of stakes need repolishing from time to time on a machine buff.

Stakes are either heavy items which can be fitted into bench sockets or held in a bench vice, or they are made as small *heads* which are held in slots in a *horse*, itself held between the jaws of a bench vice. A typical horse as shown in an illustration here, would have a number of differently shaped heads, on which various shapes could be worked.

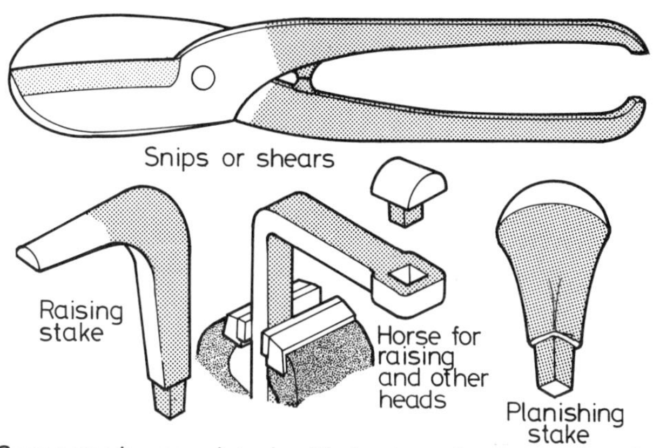

Some tools associated with beaten sheet metalwork

Three beaten metalwork articles made from gilding metal

Stages in raising sheet metal

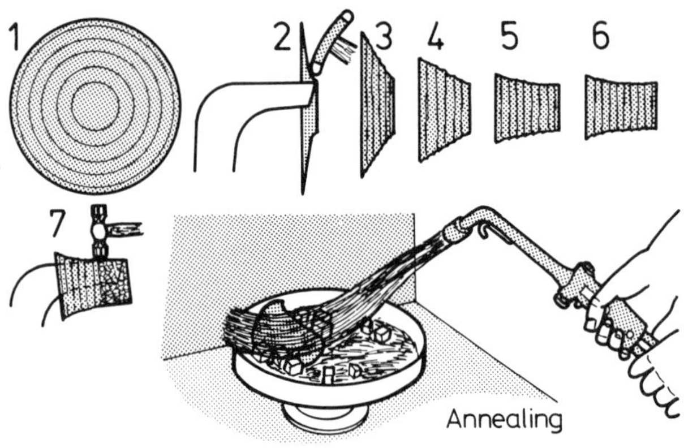

Stages in raising sheet metalwork

The Drawings 1 to 7 illustrate the stages in raising a sheet metal item such as one of the vases shown on the previous page. When metals such as copper, gilding metal and silver are worked with a hammer, they harden to become stiff and rigid. This hardening, known as *work hardening*, is when the metal becomes so rigid that it will resist further shaping. The process of softening the metal to its ductile, malleable state is known as *annealing*. Annealing of copper, gilding metal and silver is effected by heating the metal to a cherry red heat and cooling it by quenching in cold water. The metal is then placed in a dilute sulphuric acid bath (a process known as *pickling*) which removes the black oxides formed on the metal by the heating. The surfaces of the pickled items should be thoroughly cleaned with pumice powder and water, then rinsed and dried before the raising process can be resumed.

Drawing 1. The sheet metal is cut to a circular shape with shears. The base diameter of the vase is marked with a compass and pencil, and concentric circles are then drawn at about 10 mm intervals with the pencil compass. These pencil lines give a guide to the positioning of hammer blows as raising proceeds.

Drawing 2. While holding the metal disc firmly against the corner edge of a raising stake, a raising hammer is employed to make repeated blows along the first of the pencil lines as the metal is rotated by the hand not holding the hammer. Hammering is then continued on to the next pencil line, and so on until the disc edge is reached. The hammer blows must be directed at a point just above where the metal is resting on the stake in order to force the metal on to the stake. The work is then annealed, pickled and cleaned.

Drawings 3 to 6. These show the shape of the vase after the stages raising, annealing, pickling and cleaning. These stages of shaping are repeated until the desired vase form is achieved.

Drawing 7. This shows the final stage of forming the vase. After annealing and thoroughly cleaning the raised shape, a series of light blows with a planishing hammer are made, overlapping each other and covering the entire surface of the vase. The resulting finish at the end of planishing consists of a number of tiny overlapping flat and polished surfaces known as *planishing marks*. These form a pleasing patina. Planishing marks are either left on the work or they are removed by polishing.

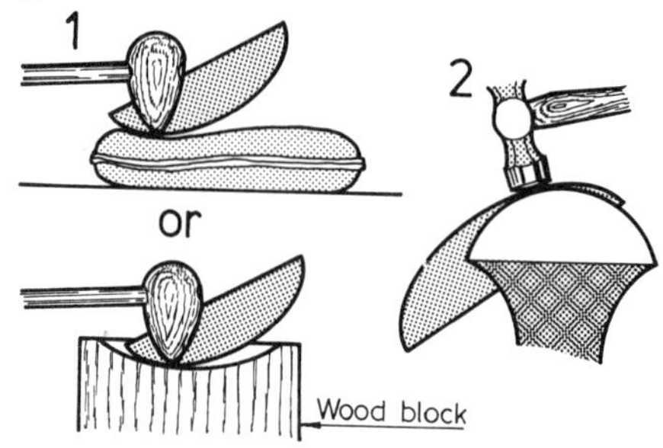

Stages in sinking sheet metalwork

Forging

Forging is the name given to the working of metals to shape with hammers while the metal is red hot. The metals are heated in a forge of which two types are common in schools – coke-fired and gas-fired. The gas-fired forges heat ceramic chips with the aid of a continuous stream of air driven through the heated chips, whereas coke forges burn coke with the aid of an air blast. Ceramic chips do not burn and therefore do not need replacing, and neither do they form ash as does burning coke.

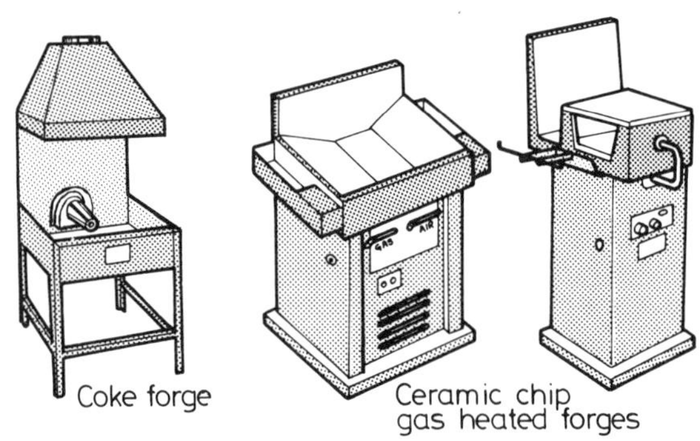

Types of forge

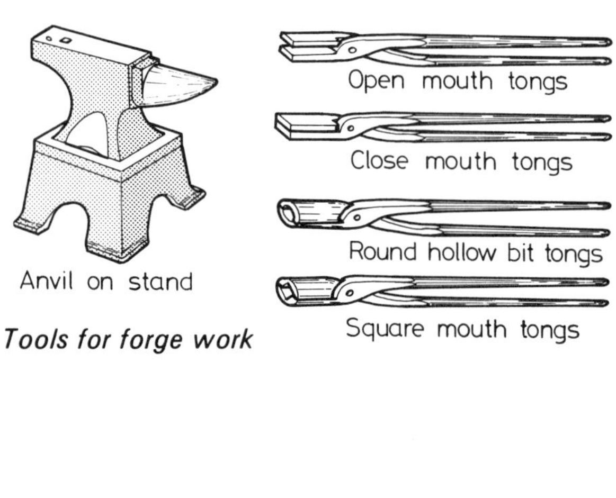

Tools for forge work

Equipment and tools for forging

Anvils

Anvils have polished steel working faces, a rounded tapered *bick* at one end, and square and round holes at the other. Most forging is carried out on the working face, whilst the bick is for working metal to a circular shape. The square hole will receive a *hardie* (see below) and the round hole enables holes to be punched in hot metal.

Tongs

Tongs are for taking metal from a forge and for holding the hot metal as it is worked. Four of the different types of tongs are illustrated on page 38 opposite.

Hammers

Engineer's hammers with a head weight of about 1 kg are suitable for most forging operations, but sledge hammers, with heads of 2 to 3 kg, should be available for heavy forging.

Other forge tools

Top and bottom *cold sett* (they form a pair) are for cutting metal to length at the anvil (see the drawing here). A hammer is used in conjunction with the top sett and the bottom sett is known as a hardie.

A number of different sizes of punch should be available for punching holes of different diameters in hot metai.

Other forging tools will be seen in the drawings.

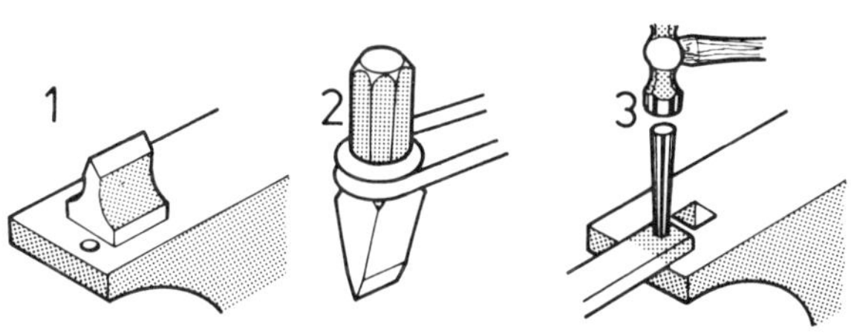

1 A cold sett or hardie. 2 Hot or cold sett. 3 Punching a hole in hot metal.

Notes on forging

1. We are really only concerned here with the forging of mild steels, which should be heated to cherry red for forging. On the rare occasions when carbon tool steels are forged, do not heat above cherry red. At a higher heat, carbon tool steel will 'burn' and emit sparks at which point the steel will be ruined.
2. Only that part of the metal being forged need be heated. Careful quenching of other parts in water may occasionally be necessary.
3. To allow them to cool, place worked pieces in a safe place, where another person is not likely to pick up a hot piece. Black hot metal does not look hot but if handled it can cause severe burns.

Examples of forging

Forging a taper

This is known as *drawing down* an end.

1. Heat the metal to cherry red and take the piece from the forge with tongs. Hammer both faces of the hot metal to draw out the taper.
2. The edges will spread as the faces are drawn down, so if necessary reheat the metal and hammer the edges parallel. The faces may again require to be worked after relieving the edge spreading.
3. The completed taper.

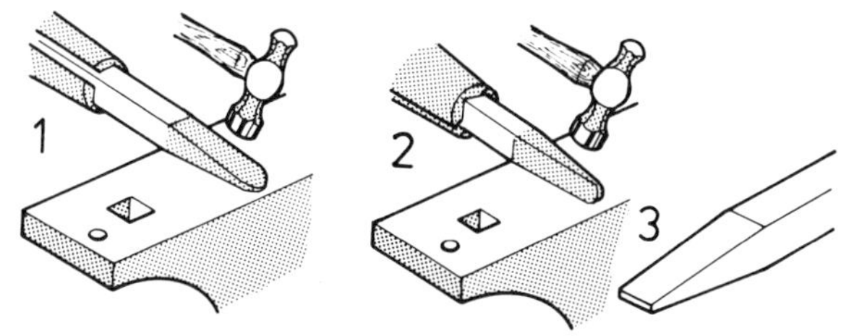

Stages in forging a taper

Forging a bend

1. Heat the metal to cherry red and remove it from the forge with tongs. Quench the red hot parts not at the bend in water and commence the bend at the edge of the anvil face.
2. Reheat if necessary and complete the bend.
3. Turn the work on its side and hammer out any spreading.

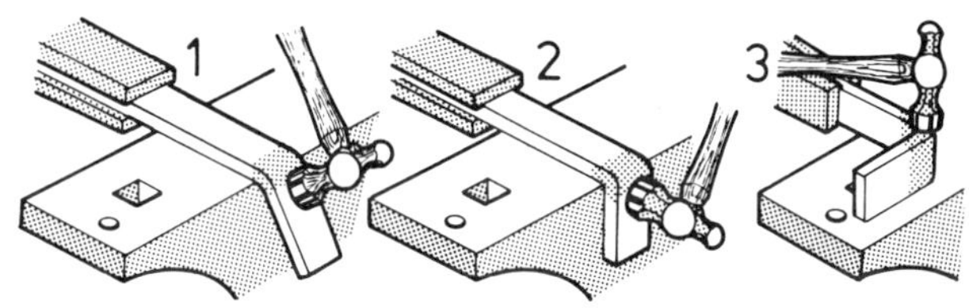

Stages in forging a bend

Forging an eye

1. Heat the metal to cherry red and remove with tongs. Quench all but the part being bent and hammer a bend at 90°. The length bent should be three times the diameter of the eye.
2. Reheat and commence forming the eye on the bick of the anvil.
3. Reheat and complete the eye on the bick.

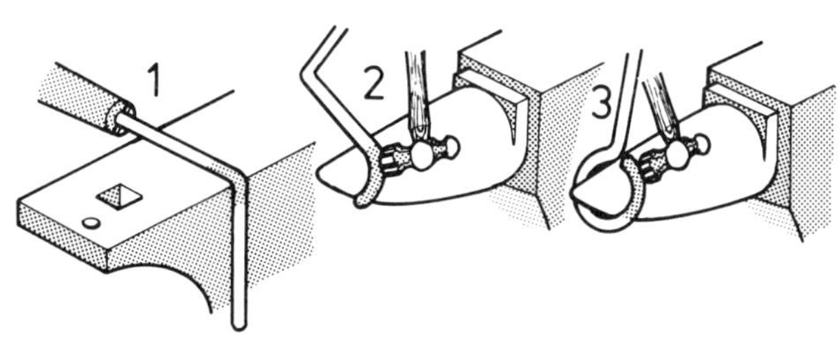

Stages in forging an eye

Forming curves in wood

There are three workshop methods of producing curved shapes in wood.

Curved shape sawn from solid wood

The shape of the part is marked out in pencil on wood of the required thickness and the shape is sawn from the board using a *bow saw, coping saw* or (for thicker wood) a *machine band saw*. Drawing 1 shows that the grain crosses the wood, and is very short across parts of the main curve. This may be a source of weakness under strain. Unless several similar parts are required, this method can be wasteful.

1

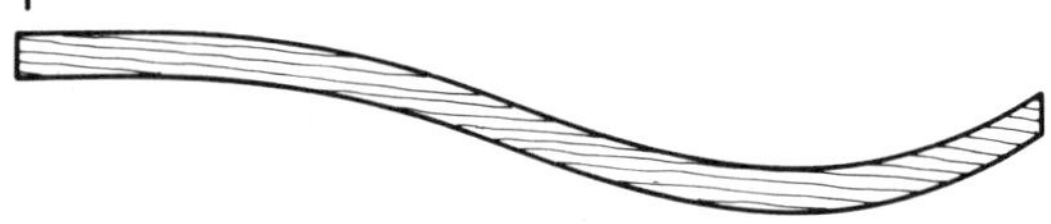

Curved shape sawn from solid wood

Curved shape built up from laminations

This method, shown in Drawings 2 and 3, will produce a curved piece of considerable strength, as the grain runs along the curve. Several thin strips of wood (*laminations*) are glued together between formers or moulds in which they are held by cramps until the glue sets hard. The shaped pieces are then removed.

Either strips of paper, or strips of sheet polythene should be placed along the inside edges of the moulds to avoid the possibility of the laminated piece becoming glued to the mould faces.

Again, unless a number of identically shaped pieces are required, laminating by this method may become expensive because of the cost of materials for the moulds and the time needed for making them.

Curved shapes formed from steamed wood

When wood is saturated with steam, it becomes pliable and, if then held in a curved shape until the wood cools, the shape becomes permanent. The resulting curved piece is extremely strong because the grain of the wood runs along the curve and not across any part of it. Drawings 4 and 5 illustrate this method.

Drawing 4 shows a school-designed steam box made from wood. A lid in the left-hand end of the box will lift off to allow lengths of wood to be placed inside and a pipe in the lid is fed to a can of water being boiled on a gas ring. (Excess steam is allowed to escape via a small hole in the other end of the box.) Wood is placed in the box, the lid closed and steam fed into the box for an hour or so to allow the wood to become sufficiently pliable to be bent. The wood is then removed from the steam box and cramped between moulds, where it is left until it has cooled, when it can be taken from between the moulds.

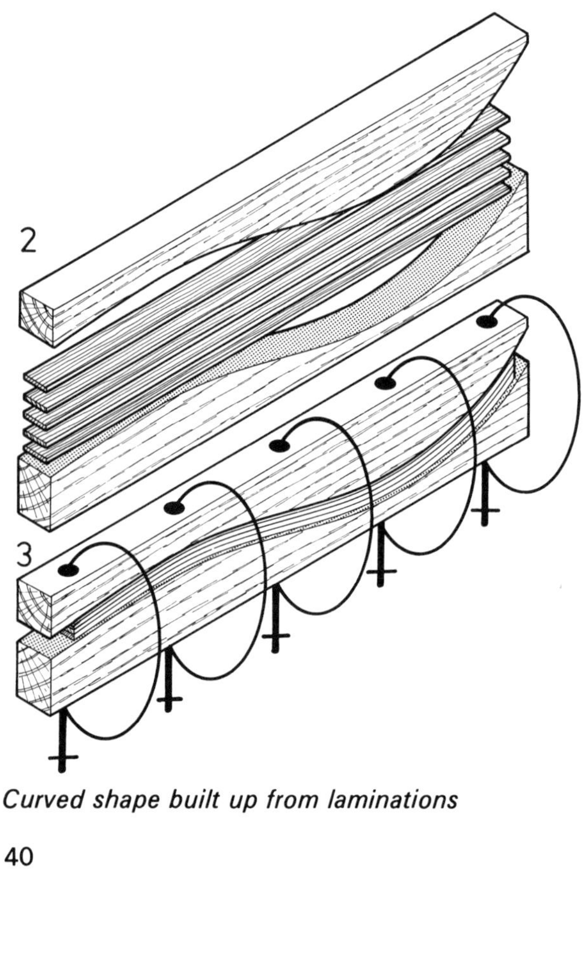

Curved shape built up from laminations

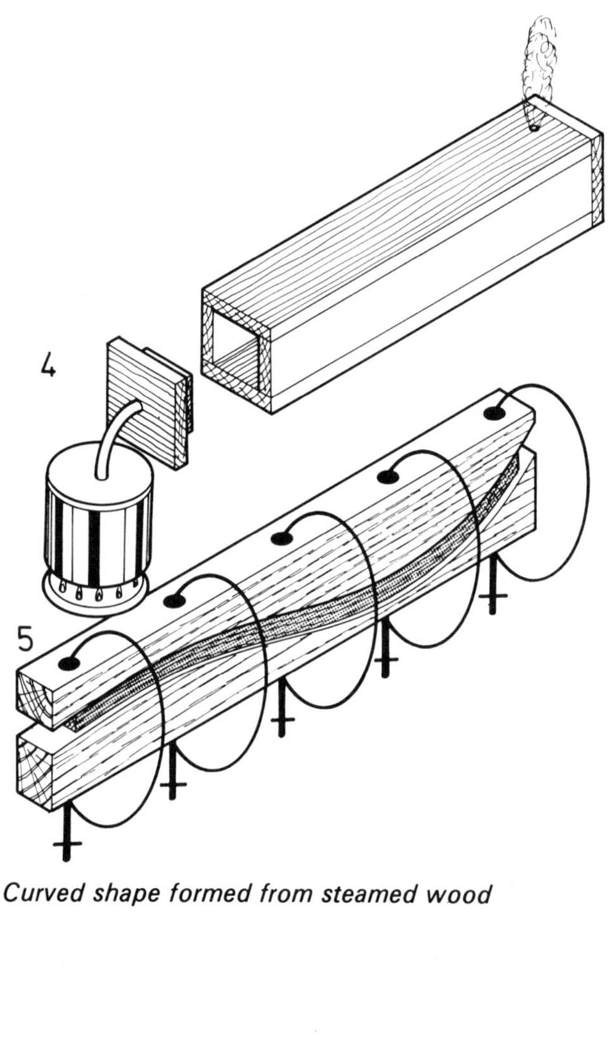

Curved shape formed from steamed wood

Heat joining of metals

Three methods of heat jointing metal parts are common in schools – *soft soldering*, *brazing* and *silver soldering*. A fourth method, *welding*, can only be carried out under authorised instructors or teachers. Hard soldered (brazed and silver soldered) and welded joints are very strong, and soft soldered joints are relatively weak.

Soft soldering

Soft solders
Soft solders are alloys of tin and lead plus a small percentage of antimony (1% to 2%) and they melt in the range from about 180° C to 240° C. Solders for fine work such as electrical and electronics circuitry have more tin than lead (65% tin) while those for general work may have up to 65% lead.

Fluxes for soft soldering
As metals are heated, oxides and other impurities rapidly form on their surfaces. Such impurities prevent a joint from being soldered. *Fluxes* are for either chemically cleaning the metal surfaces which are to be soldered, or for acting as barriers to prevent air from reaching the surfaces being soldered. Only three soft solder fluxes are mentioned here – the proprietory 'Baker's' fluid, zinc chloride (made by dissolving zinc in hydrochloric acid) and resins (which act as barriers).

Soldering irons
Large soldering irons are usually heated in a gas fired soldering stove whereas smaller types are electrically heated. The *bits* of soldering irons are made of copper because copper has excellent heat-conducting properties. The tips of the bits must be kept *tinned* as follows: file the tip clean; dip it in flux; then coat the tip with solder while hot.

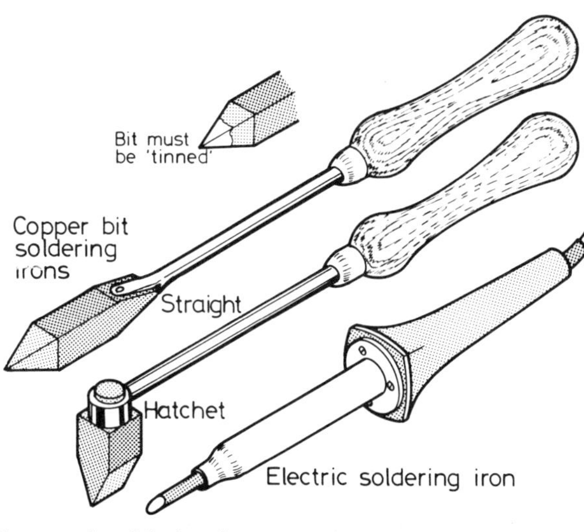

Types of soldering iron

Stages in making a seamed, sweated joint
1. When the flames from the soldering stove turn green, the bit of the soldering iron will be at a suitable temperature for soldering. Overheating damages the bit.
2. Apply flux to both surfaces to be joined and dip the hot bit in flux to clean it. Then pick up solder on the bit and coat the surfaces to be joined with solder.
3. Place the two parts together under pressure from a scrap piece of metal (NOT a file) then heat the joint until the solder melts. This is known as *sweating*. Allow the joint to cool while under pressure.

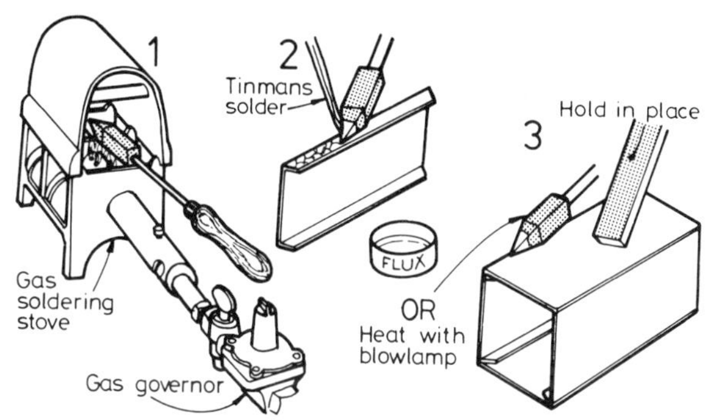

Stages in making a seamed sweated joint

Stages in soldering a wire lead to a transistor
The pliers, held to the transistor lead by rubber bands, act as a *heat sink* preventing the heat from the soldering iron damaging the transistor.

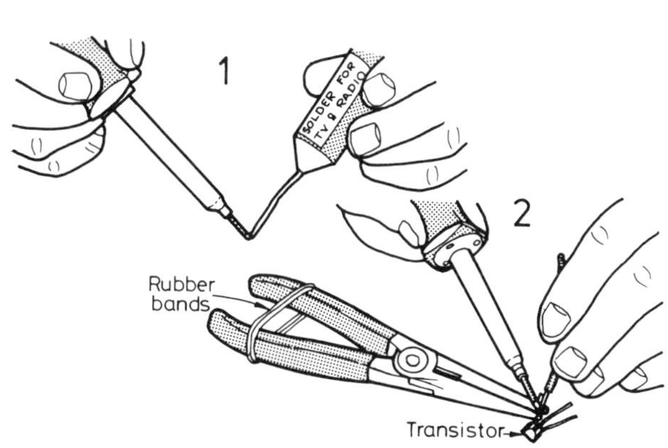

Stages in soldering a lead to a transistor using pliers as a heat sink

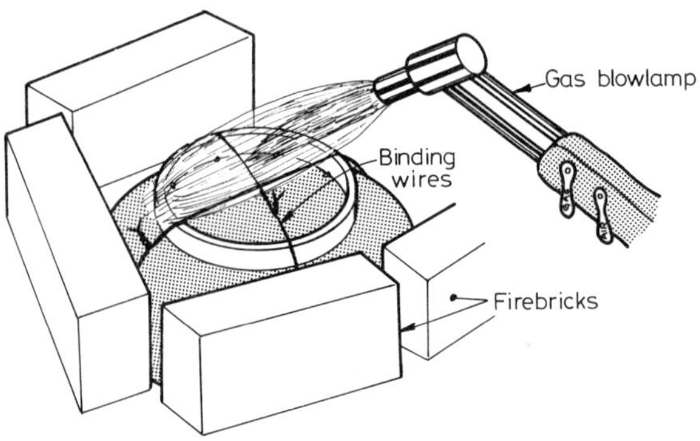

Soldering a base on a bowl with the heat from a brazing torch

Soldering a base to a bowl

Gentle heat from a blowlamp will heat small pieces of soft solder into a cleaned and fluxed joint.

Hard soldering

In this book the term hard soldering is used to include both brazing, in which brasses are melted into a join, and silver soldering in which alloys of silver are melted in a join. Both forms of soldering produce strong joins, and both involve similar methods of working.

Brazing spelter

Brazing 'spelters' are brasses (alloys of copper and zinc) used for brazing. They are made in either wire form or as granules, and several types are available. A suitable spelter for mild steel would be a copper/zinc alloy in the ratio 2 : 1. Spelters for non-ferrous metals, such as copper and brass, would contain more zinc, with copper/zinc alloys in ratios of 3 : 2 or 1 : 1. The melting points of these spelters are between about 850° and 900° C; those with more copper are in the higher melting range.

Silver solders

These are alloys of mainly silver, copper and zinc, with the proportion of silver varying from as little as 50 % in the case of those melting at around 625° C, to as much as 80 % in solders melting at around 800° C. The range of silver solders embraces several grades, from *Easy-flo* which melts at the lowest temperatures and is very quick running, to enamelling solders for work which must later be enamelled. These solders melt at the higher temperature of 800° C.

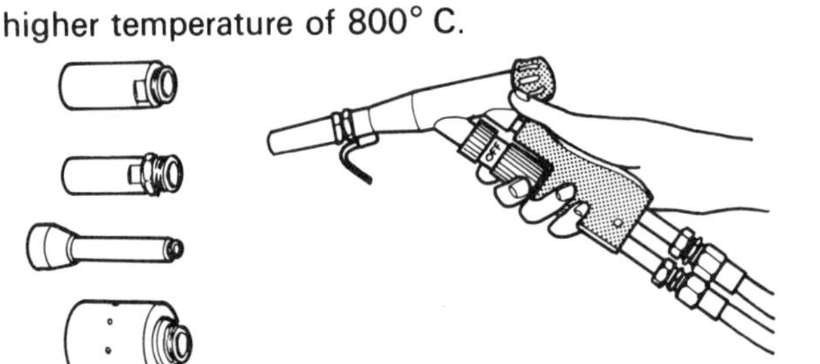

A modern blowlamp operated by gas and air, with spare torches

Blowlamps

The heat required to melt brazing and silver solders into their joins is usually provided by means of gas/air blowlamps which have gas and air controls to regulate both the size and the heat of their flames. Such blowlamps are often fitted with interchangeable nozzles which increase the range of flame size which can be produced. Blowlamps are used in conjunction with some form of brazing hearth where the work can be held in refractory, heat-resisting bricks or in cubes on a heat-resisting background. Some hearths are made to revolve so as to allow heat to be applied from all sides of the work.

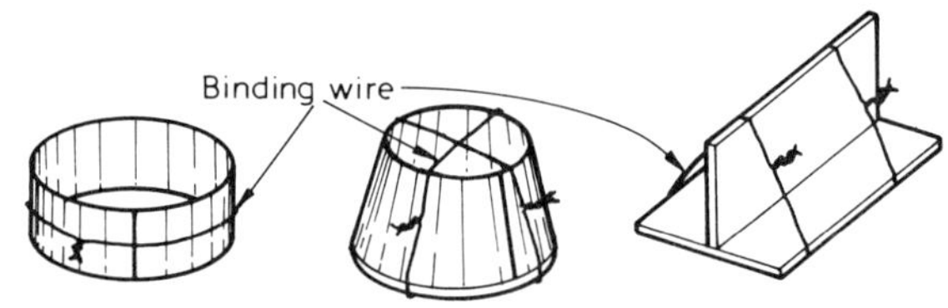

Temporary joining with binding wire prior to brazing or silver soldering

Fluxes

The common flux for hard soldering is *borax*. This is mixed with water to form a thick paste for application to parts of joints. A variety of proprietory fluxes are also available and it is advisable to use the correct proprietory flux when using solders such as Easy-flo. Before fluxes can be applied, the parts of a join must be thoroughly cleaned as any dirt or impurity within a join will result in the join failing to make. Borax melts at a lower temperature than the solder and flows into the joint excluding air and also dissolving unwanted oxides formed during the heating process.

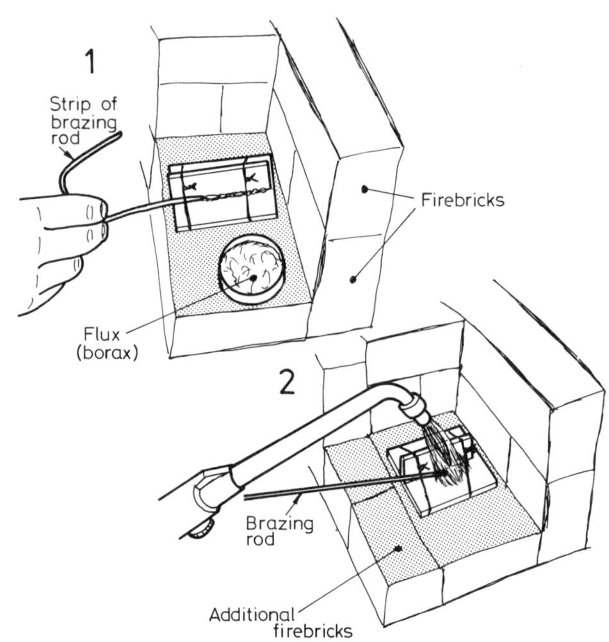

Stages in brazing or silver soldering

Stages in hard soldering
The processes involved in both forms of hard soldering are the same.
1. Thoroughly clean the parts of the joint, down to bare metal.
2. Coat the joining areas with a paste of flux.
3. Ensure that the parts are held together in such a manner that they will not slip under the action of the blowlamp's flame. One way of ensuring this is to bind the parts together with iron binding wire. Another is to place them in the brazing hearth and pack them all round with firebricks or heat-resistant cubes in such a manner that the parts will not slip against each other.
4. *Drawing 1* (Page 42.) Place the temporarily assembled pieces in a brazing hearth and apply flux along the joint. This can be done with a strip of brazing rod or brazing wire or with a strip of silver solder.
5. *Drawing 2.* Now bring all the work to red heat with the flame of a brazing torch. Then apply the end of the spelter or silver solder to the joint and, when the heat is right, the spelter or solder will melt and flow into the join.

Welding

Safety precautions
All welding in schools must be carried out only under the close supervision of instructors and teachers who are themselves qualified in welding practice. It is of particular importance that operators wear goggles fitted with filter lenses to protect the eyes against the light emitted by the welding arc or welding flame, as brief exposure to the light emitted by welding can cause permanent damage to the eyes. Welding should be carried out in special booths or in places where others not engaged in the work will not be affected by the light emitted by the welding. The body of the operator should also be protected, preferably by wearing welding overalls, and the hands are liable to be burnt by hot metal spattering back from the work, so that it is advisable to wear gloves. Note that the intensity of light from electric arc welding is greater, and of a different form, to that emitted from oxy-acetylene welding and requires greater eye and body protection.

Types of welding equipment
To achieve the high temperatures necessary to melt the work when welding, two types of welding equipment are manufactured – electric arc welding units, and oxy-acetylene welding equipment. The power source for the electric welding units in use in schools is usually mains electricity, to which the units are fitted. These units include transformers, which reduce mains voltage to between 50 and 90 Volts. (This voltage is further reduced once an arc has been struck.) In oxy-acetylene equipment the heat is provided by the flame produced by burning acetylene in oxygen. The two gases are obtained from high pressure gas cylinders. In this book we are concerned only with oxy-acetylene welding.

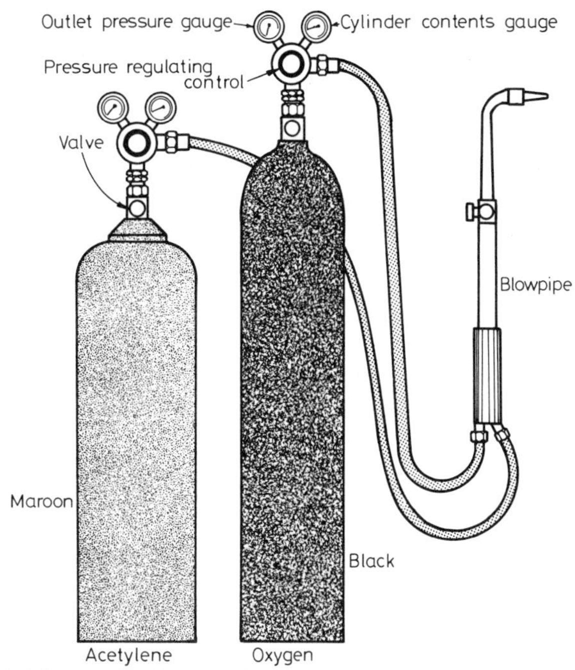

Welding cylinders and a welding blowpipe

Gas cylinders
Oxygen is contained in solid-drawn alloy steel cylinders under pressure. These cylinders are taller and narrower than acetylene cylinders and the valve outlets are made with right-hand screw threads.

Acetylene cylinders are short and wide in comparison, with valve outlets which have left-hand screw threads. A porous mass of charcoal or of lime silica, in which liquid acetone is held, is placed in the cylinders. Acetone has the ability to dissolve up to 25 times its own volume of acetylene at normal barometric pressure, and under pressure in the cylinders, considerably more than this amount of acetylene is dissolved. The volume of acetylene which can be held in these cylinders is thus increased substantially.

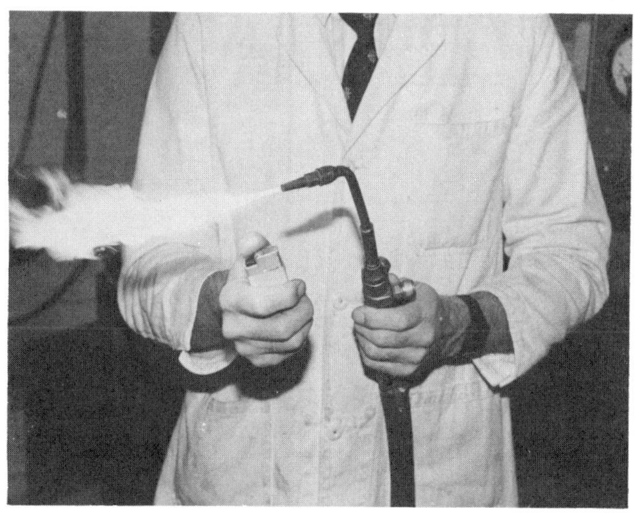
Igniting the acetylene

Welding blowpipes

Blowpipes are attached to the gas cylinders via colour-coded hoses – red for acetylene; blue for oxygen. As discussed above, hose connections to the oxygen cylinders are made with right-hand screw threads and hose connections to the acetylene cylinders are made with left-hand screw threads. The hoses are connected to the blowpipe again with right-hand threads for the oxygen connection, and with left-hand threads for the acetylene connection. Blowpipes are designed so that the gases are mixed in a mixing chamber within the blowpipe and burning takes place at the blowpipe nozzle. Welding blowpipes are made with detachable interchangeable nozzles of different sizes to allow the best flame to be used for the welding of different thicknesses of metal.

Stages in welding

1. Check that the cylinders contain sufficient gas, and before lighting up, turn on each valve for both gases in turn to ensure that gas pressures as shown by the gas gauges are correct. Then close both gas valves on the blowpipe.
2. Slowly open up the acetylene valve on the blowpipe and light the gas. Then adjust the flame until it stops smoking.
3. Gradually open the oxygen valve on the blowpipe and adjust it to produce a welding flame. This is recognisable by a sharply defined white or luminous inner cone to the flame. A welding flame is produced when about equal quantities of the two gases are being mixed within the blowlamp.
4. If possible *tack* the ends of the parts to be joined.
5. Welding can now proceed.

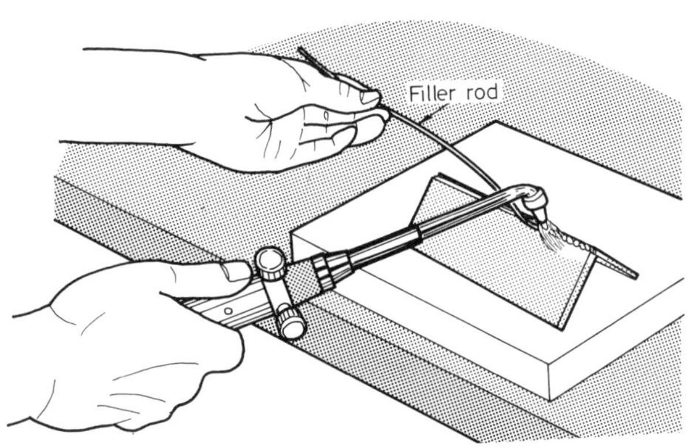

Forming a welded joint

Note: the above description of the welding process is only a bare outline. Welding is by itself a complete craft requiring great skill from its operators. Further information from instructors or teachers themselves qualified in welding practice should be obtained before attempting to make welded joints.

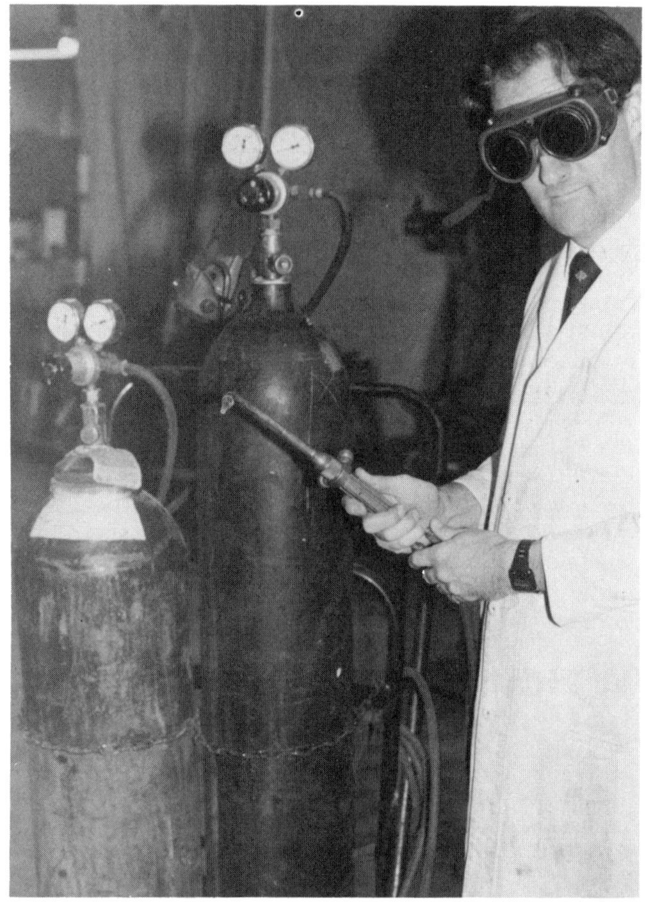

Acetylene and oxygen cylinders and welding blowpipe. Note: safety goggles

Tacking the ends of a joint ready for welding

Encapsulation in polyester resins

The enclosing of articles within clear polyester resins is a simple operation and very good results can be achieved if care is taken when making the encapsulation. The encapsulation may be for decorative purposes – brooches, rings, pendants, paper weights and lamp bases being examples – or it may be for the preservation of biological specimens such as insects, butterflies, moths, small skeletons, flowers and seed heads. Polyester resins for encasing such items are sold as clear casting resins and they are formulated especially for the purpose.

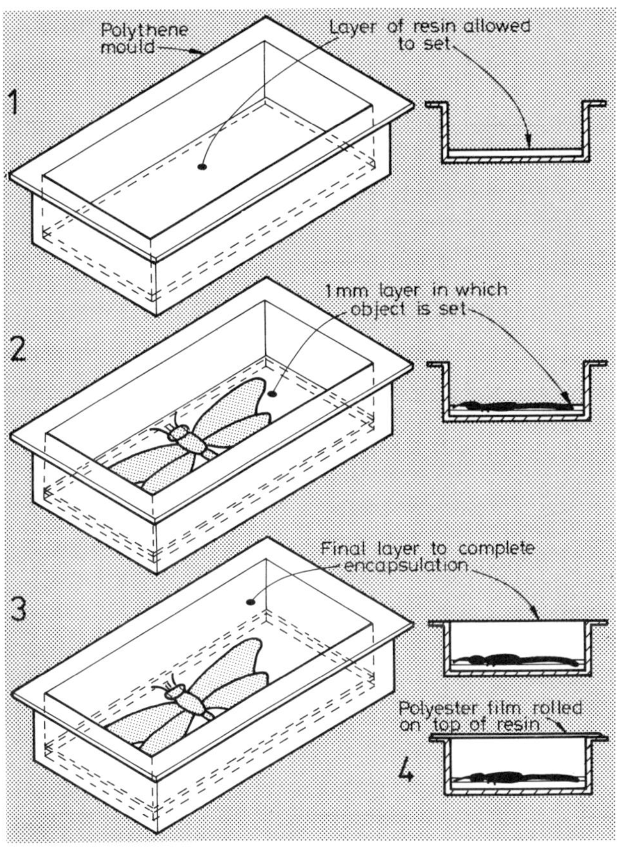

Stages in making an encapsulation with polyester resin

Making an encapsulation

Stages in making an encapsulation

Stage 1
Mix sufficient clear casting resin with 2% of catalyst to form a layer about 6 mm deep in the bottom of a polythene casting mould. Pour the catalysed resin into the mould, and put the mould to one side with a piece of cardboard covering the mouth of the box. (The cardboard will prevent dust settling on the surface of the resin.) Any air bubbles in the resin should rise to the surface and break.

Stage 2
Allow the resin to partially gel – this will take perhaps two hours because of the low activator content in clear casting resins. At this partial gel stage carefully place the item which is to be encapsulated on to the resin and again put the mould to one side with its cardboard cover replaced for several hours until the resin sets – overnight would be a convenient length of time.

Stage 3
Mix up a sufficient quantity of clear casting resin with 2% of catalyst to fill the mould and pour enough of the catalysed resin carefully on to the item being encapsulated, to completely fill the mould. Leave the resin to gel – which will be perhaps two hours after pouring. At the gel stage roll a piece of polyester film on to the top surface to allow the upper surface to polymerise fully in the absence of air caused by contact with the polyester film. Allow the casting to remain in its box for a further two days, when the polyester film can be peeled from the top. The encapsulation should now come free from the mould – a gentle tap or two on the outside of the mould may be necessary.

If the polyester film has been laid without trapping air bubbles, then the upper surface of the encapsulation should be smooth and glass-like. If it is not smooth, then the surface will have to be sanded and polished.

The layers of resin can be seen but are not obtrusive.

A butterfly, a dragon-fly and a large centipede encapsulated in polyester resin

Aluminium Sand Casting

Stages in making an aluminium casting in sand are described. The photographs and text are taken (with permission) from a wall chart *Aluminium Sand Casting* published by William Allday and Co. Ltd., who are the makers of the Alcosa range of metalworking equipment. A warning printed on the poster states 'This poster should be used as a guide only – full instructions will be found in the Alcosa *Aluminium Casting* booklet'. The booklet is also published by the firm.

1. Sand must be of the right consistency, this is very important. It should be free of lumps and well mixed. To check moisture content, squeeze a handful into a lump. When broken apart it should separate into two pieces without crumbling.

2. Assuming the sand is in the right condition, casting can proceed. First a double box, called a flask is needed, in which to form the sand bed. The upper half is called the cope and the lower half the drag.

3. Place the drag upside down on a turnover board. Place the pattern near the centre, detail side uppermost. Leave at least 25 mm or more between the flask and the pattern.
Sprinkle the pattern and the turnover board with parting powder shaken from a linen bag.

4. Riddle sand over the pattern through a fine sieve mesh to approximately 25 mm deep.
Fill round the edges of the drag with unsieved sand.
Rain sand lightly but firmly, using finger tips to press into every detail of the pattern from edge to centre.
Fill the drag with unsieved sand and use a rammer to compact it during filling.

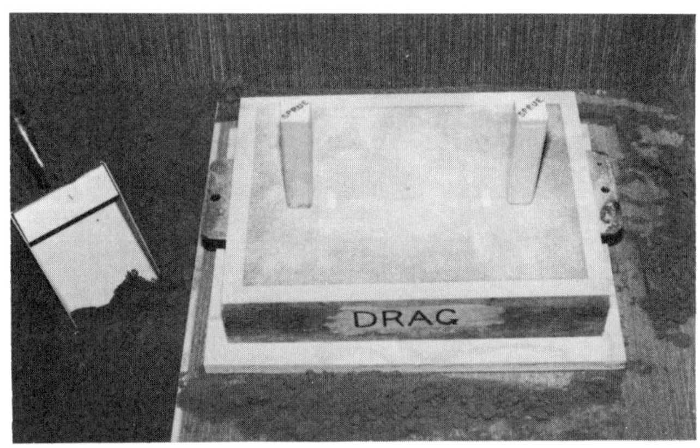

5. Turn the drag the right way up using a second turnover board on top of the drag.
Remove the original turnover board and dust the upper surface of the drag and the pattern with parting powder.
Press sprue pins into the sand each side of the pattern – positioned near the bulkiest parts. Fit the cope on top of the drag then proceed as in 4 above.

6. Scoop out a pouring basin next to one sprue pin, forming the runner into which metal is poured. The other sprue pin forms the riser.
Remove the sprue pins and make a few venting holes in the cope with a thin metal rod stopping approximately 6 mm from the pattern.
Lift the cope from the drag and carefully set it aside.

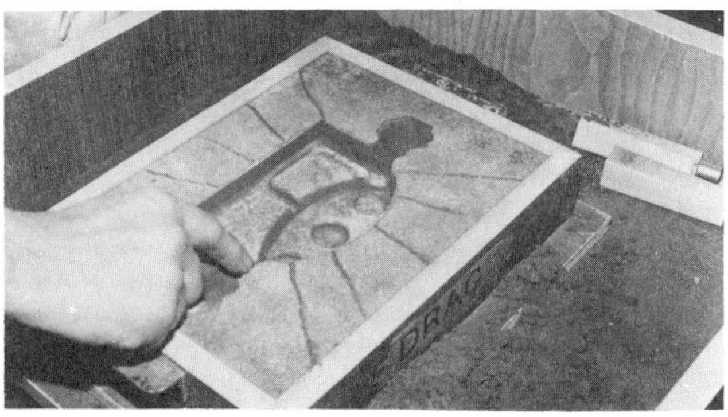

7. With the pattern still in place cut horizontal channels from the pattern to the sprue holes, large enough to allow free flow of the molten metal.
Scoop out a depression where the runner sprue rested, to provide a trap for sand particles washed down.
Make a few shallow scratches from the pattern to the edge of drag.
Carefully remove the pattern from the sand, rub off any sharp edges in the sand channels and blow away any loose sand particles.
Replace the cope on the drag. Place the flask on a sand bed near to the furnace ready for pouring.

8. Bring the furnace to its working temperature, at the same time pre-heating the metal ingots by placing them on or around the furnace lid. Using tongs place the pre-heated ingots in the crucible.
When the metal begins to melt sprinkle a small amount of flux on to the surface of the metal to form a slag and prevent oxidation.

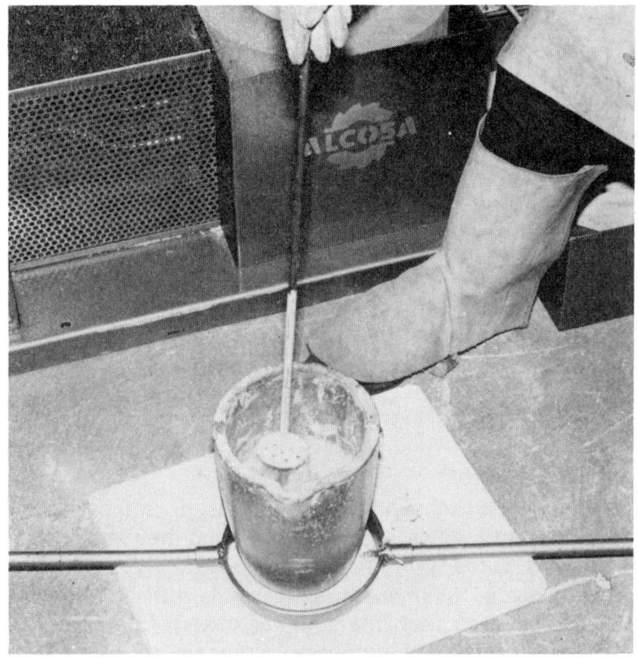

9. When the metal is at its pouring temperature (600° to 630° C), remove the crucible from the furnace using pre-heated tongs and place it into a pouring ring on a sand bed.
Skim off the slag from the surface of the metal. Drop in a tablet of de-gassing agent and plunge it slowly to the bottom of the crucible (especially important if scrap metal is used).

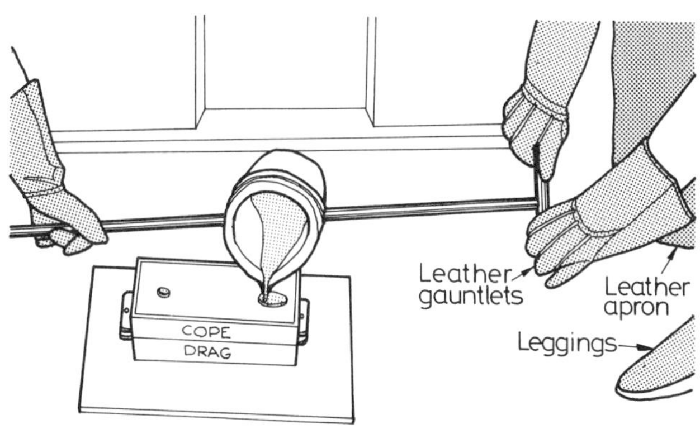

10. Carefully pour the metal into the runner until it appears at the top of the riser. Any surplus metal should be poured into pre-heated ingot moulds. When the casting has solidified the mould may be broken up and the riser cut off.

Glass reinforced plastic (GRP) mouldings

Moulds
The most common method of producing a GRP moulding is to build up layers of polyester resins and glass fibre on to moulds, with the aid of brushes and rollers. The resins are then allowed to harden on the mould. The mould may be *male*, where the moulding is built up on the outer surfaces, or *female* in which the moulding is laid on the inside surfaces.

Moulds can be made from practically any solid materials at hand. The materials most frequently employed are wood, plaster, clay, fibreglass and various moulding rubbers. As the surfaces of a GRP moulding are an exact copy of the surfaces of the mould, some care is necessary when finishing the surfaces of moulds.

Parting agents
Polyester resins have a strong affinity for many other materials and will adhere to them when the resins have set and in some work, the resins are actually employed as adhesives. It polyester resins are laid directly on to the surfaces of such materials as wood, fibreglass or plaster, then the resulting mouldings cannot be parted from the moulds because they will have become glued to them. There are some materials to which polyester resins will not adhere – some examples are polythene, paraffin wax and various moulding rubbers.

Materials to which the resins will not adhere are employed as *parting agents* between the mould and the moulding. Two common parting agents are wax polish, which must not contain silicones, and polyvinyl alcohol (pva), which is often supplied in liquid form.

Tools
Few tools are required and brushes and rollers are the most important. Purpose-made brushes for polyester resins are best, as the bristles of ordinary paint brushes may work loose under the action of the resins. Flat rollers are useless but purpose-made serrated rollers, or rollers made from mohair or lambswool, can be used.

Sharp cutting knives are necessary and a palette knife is of value for mixing colours or for parting a moulding from its mould. A pair of scissors should be at hand for cutting glass fibre.

Colours and other finishes
Resins can be coloured with pigments, or mixed with suitable powders to give a different surface appearance. Pigments can be either opaque or translucent – and metal powders – copper, bronze or aluminium – can be used to simulate metal surface finishes. Other powders, such as marble, can also be mixed with the resins.

Types of resin
Two types of polyester resin are used when making a GRP moulding. These are *gel-coat* resin and *lay-up* resin. Gel-coat resin is purchased as a slightly thickened liquid for direct application to a mould and it is the first coat of resin applied when making a GRP moulding. Lay-up resin is the general purpose resin for laminating glass fibre within a GRP moulding.

Polyester resins are purchased in a pre-activated form. In order to make them set solid, small quantities of *catalyst* (hardener) are added to the resins. Catalysts are usually peroxides, available in liquid or paste form.

Glass fibre
Although several different forms of glass fibre are available, GRP mouldings are usually built up from *chopped strand mat*. This consists of strands of glass fibre, each about 50 mm long, placed randomly to form a mat. The mat of strands is bonded together with a bonding agent which dissolves in polyester resin.

A second type of glass fibre, *surfacing tissue*, is often used to supply a reasonably smooth surface on the outside of a moulding. Surfacing tissue is a thin sheet of glass fibre which looks like thick tissue paper.

Stages in making a GRP moulding

The six drawings here show stages in making a GRP moulding which is to be used as a guard to a machine. The layers of glass fibre are built up on a mould which is itself a GRP moulding. The mould is a male mould so that the inside surfaces of the resulting GRP moulding will be smooth.

1

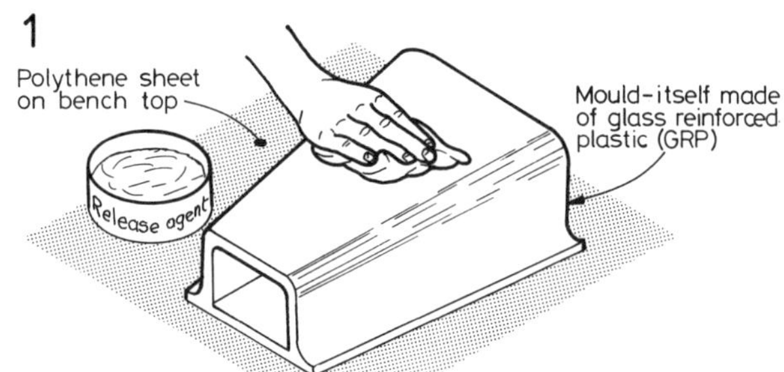

Apply release agent

1. The mould surface is polished with a silicone-free, wax polish, which will form a release agent on the surfaces of the mould. Care is taken to make quite certain that all the mould surfaces are covered with the release agent.

2

Brush on gel coat

2. Catalysed gel-coat resin is applied with a brush to a thickness of about 1 mm, using a purpose-made resin brush about 25 mm wide. (The brush is cleaned in resin cleaners after use.) The resin is allowed to gel – which takes about 30 minutes.

3

Lay-up glass fibre

3. Three layers of glass fibre chopped strand mat are applied – each layer being wetted out with catalysed lay-up polyester resin applied by brush with a vertical stippling action.

4

Roll with serrated roller

4. Another layer of fibreglass is laid up, this time using surfacing tissue with lay-up polyester resin. The surfaces are rolled smooth and compact with a serrated roller.

5

Cut off excess at 'green' stage

5. After about 30 to 45 minutes, the laid-up fibreglass is easily cut away from the mould with a sharp knife. This is the *green stage*; too soon and the fibreglass will fracture; too late and the fibreglass will need to be sawn.

6

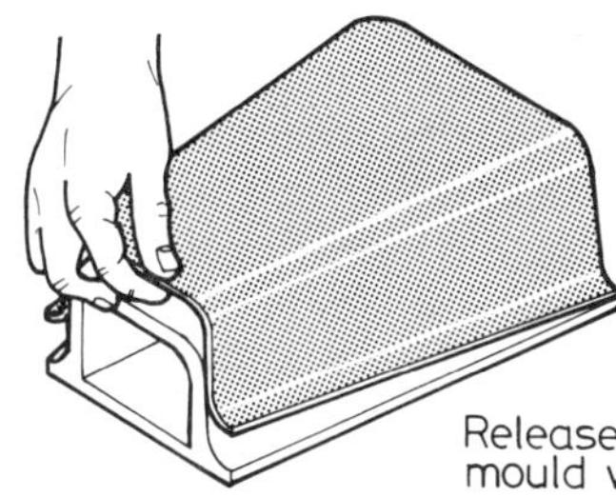

Release from mould when set

6. After a further period of four or five hours, the moulding can be parted from its mould. A palette knife inserted between the mould and moulding can be of assistance for parting the moulding.

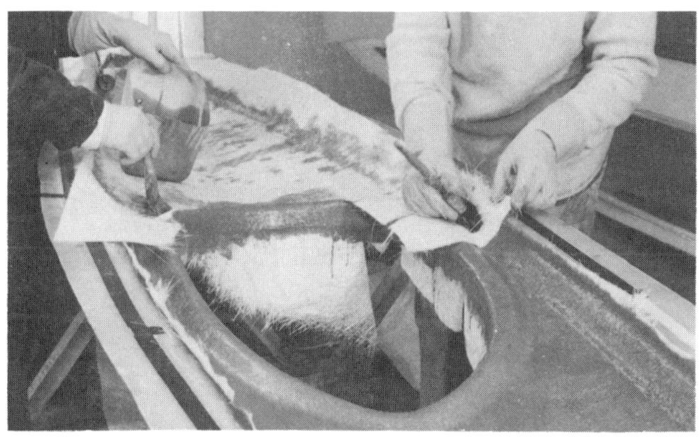

Laying-up glass fibre while making a GRP canoe

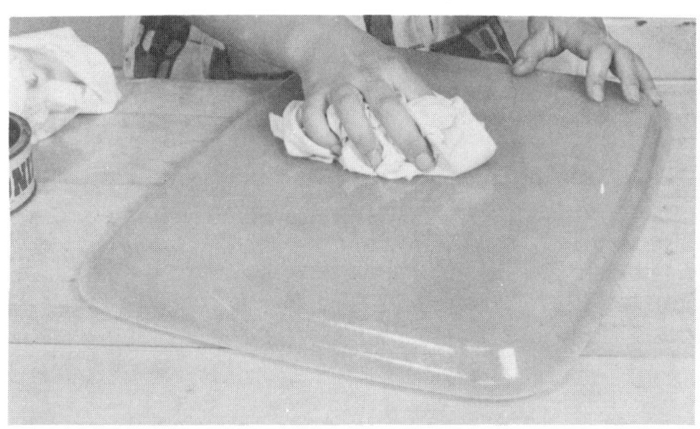

Applying release agent to a tea-tray fibreglass mould

Finishing

The surfaces of some metals and plastics used for craft design may not need protective finishes to be applied after shaping, forming and fitting together the parts of the design. Some metals such as copper, silver, aluminium, stainless steel, may be either resistant to staining and corrosion or they may need only occasional polishing with metal polish. In some situations indeed, some metals when left to *weather* may be better left unfinished, an example being the pleasing green patina which forms on copper roofs. Many plastic surfaces of vinyls, acrylics, nylons and fibreglass can be left in their original crafted state without any form of surface protection being applied. In fact some plastic materials are deliberately applied as a *finish*, an example being the addition of plastic laminate sheet to the surfaces of kitchen furniture.

Most materials used in school craft and technology designs will, however, need the surfaces of the materials to be worked smooth and, in addition, many will require some form of protective finish to be applied in the form of stains, paints, varnishes, lacquers or polishes.

Smoothing wood surfaces

To obtain a smooth and clean surface on wood, ready for the application of a surface finish, the following procedure should be followed:

1. Plane the surface flat and smooth with a sharp, very finely set smoothing plane. The cross-grained surfaces of some hardwoods may need to be planed with a sharp, very finely set block plane.
2. Any defects remaining on the surface of a hardwood after such planing can be scraped out with a *cabinet scraper* (see below). Cabinet scrapers are not effective on softwoods and can indeed cause a smooth planed softwood surface to become rougher.
3. Finally, work along the grain of the wood with a sandpaper wrapped around a sanding block.

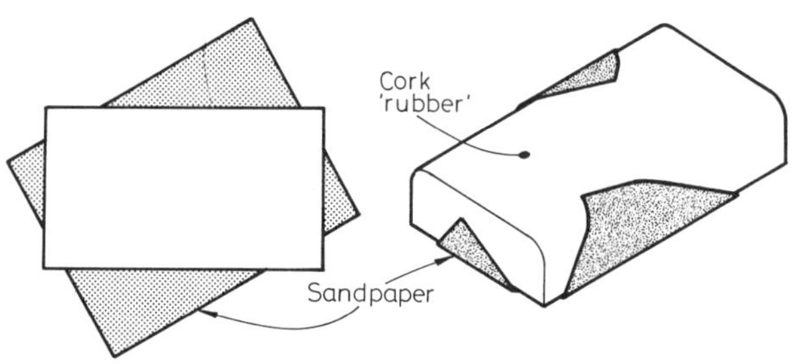

Wrapping sandpaper around a cork block

Cabinet scraper

A cabinet scraper is a flat, rectangular piece of carbon tool steel tempered to about the same hardness as a handsaw blade. The edges of the steel are filed straight, then polished on an oilstone and burred over with the aid of the rounded back of a gouge. The burred edges will scrape thin shavings from the surface of hardwoods.

Sandpapers

Four types of abrasive materials are glued to the surfaces of tough papers to make *sandpapers*. These are – glass, garnet, aluminium oxide and silicon carbide. Glasspaper produces a very smooth fine surface, but wears quickly; garnet paper will not produce as smooth a surface as will glasspaper, but it is tougher and it will not wear out so quickly; aluminium oxide paper is very tough and long lasting, but it is more often used in sanding machines than for hand work; silicon carbide paper is the *wet and dry* paper made for rubbing down old paintwork to a smooth finish (where the paper is wetted with water, or with white spirit, to prevent the dust from the paint from becoming a breathing hazard).

When any of these abrasive papers are used for hand-work, they should be wrapped around a sanding block, preferably one made from cork, as they wear out quickly if not supported by some form of block as they are used. If a cork block is not available, a block of wood is a reasonable substitute and for the final smoothing of curved edges, the sandpaper can be wrapped around a file, a dowel or a shaped piece of wood.

When sanding a surface smooth with sandpapers start with a medium grade and finish with a finer grade. Thus, if using glasspaper, grade M2 followed by $1\frac{1}{2}$ is suitable and when using garnet paper a good finish will result from grade $\frac{1}{2}$ followed by grade 2/0.

Smoothing metal surfaces

Many metal surfaces will not need to be worked smooth, particularly if the work is made from sheet materials such as tinplate, sheet copper or sheet aluminium. When the surfaces of metal do need to be worked to a smooth finish, then the following procedure can be adopted:

1. File the surface to as smooth a finish as possible with a smooth-cut file.
2. Draw-file the surface with a smooth-cut file. Occasionally work chalk into the file teeth to prevent pieces of the metal being filed being caught within the

teeth and scoring the surface (this is known as *pinning*).

3. Wrap emery cloth around the file and work it along the surface under hand pressure until the surface is smooth. Several grades of emery cloth may be needed, commencing with a medium or a coarse grade and going on to finer grades. If the draw-filing has resulted in a reasonably smooth surface, then start with grade F emery cloth and finish with grade O. Otherwise commence with grade 1, before using grades F and O. It is not advisable to commence with too fine a grade of emery cloth because the effort put into achieving a good finish will be unnecessarily prolonged and there will be considerable wastage of emery cloth. When using emery cloth a few drops of oil on the metal surface from time to time will minimise the risk of scoring the polished surface.

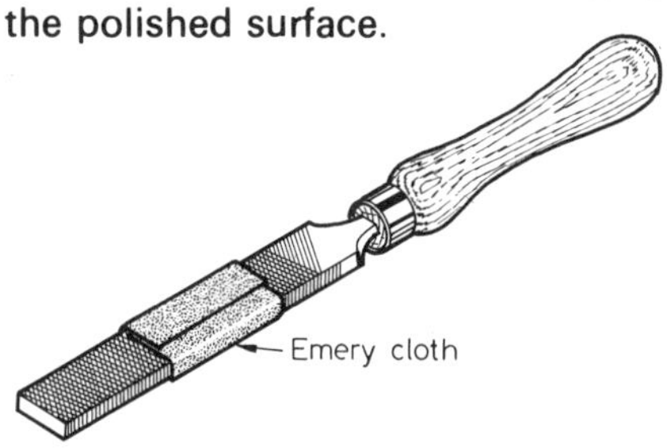

Wrapping emery cloth around a file

4. If a highly polished surface is required, the metal can be further polished on a *machine buff*. Buffs are polishing mops mainly made from a number of calico discs held together between washers. They are fitted on to the tapered screw at the end of the spindle of a grinding machine. A bar of polishing compound is pressed against the edge of the buff as it is revolved in the machine. The metal surface is then pressed against the revolving buff to be polished under the action of the compound on the buff's edges. The metal is moved around so that the whole of its surface is polished by the buff.

5. Metal polishes can also be used on metal surfaces to obtain a very smooth polished surface.

Other finishes

Brushes

Paints, varnishes, lacquers and polishes may be applied to the surfaces of most materials with brushes. The type of brush most often used in hand craft work is the *flat varnish brush*, made in widths of 12 mm, 25 mm, 50 mm and 75 mm, the most useful of these being the 25 mm.

Brushes must be cleaned after use. The fluids in which they should be cleaned will vary according to the finish which has been applied. Some examples are:
most paints – white spirit or turpentine;
cellulose lacquers – cellulose thinners;
french polish – methylated spirits;
polyurethane varnish – white spirit.

After cleaning the brush in an appropriate fluid, it should be thoroughly washed with soap and hot water to remove all traces of both the finish and the cleaning fluid. Then place the brush to one side to dry.

Dip-coating of metals

Low density polythene powder, available in a variety of colours, can be melted on to metal surfaces. This coating colours the metal surface, renders it waterproof and provides good electrical insulation.

The surfaces of metals to be dip-coated must be thoroughly cleaned and free from scale, oil or grease. All sharp edges should be removed.

The article to be coated needs to be heated to about 300° C, preferably by being hung in an oven. Alternatively, it can be heated by the flame of a blowlamp but the resulting temperature will be only a crude approximation to 300° C. The article can be held by suspending it from a wire in pliers.

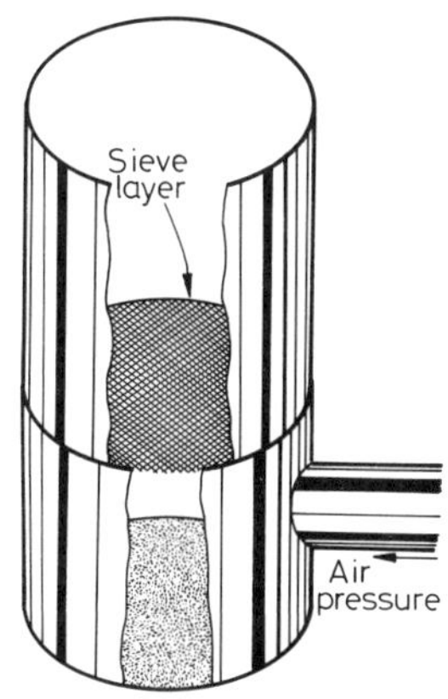

Diagrammatic drawing of a fluidising tank for dip-coating

Fluidising tank

To obtain the best results from dip coating, producing an even thickness of plastic skin over the metal, the heated metal should be dipped into powder through which air is blown. Fluidising tanks for this purpose can be purchased ready-made, but they can also be made in the school workshop. The procedure for using a fluidising tank is outlined below.

Place some polythene powder in the upper container, remembering that when an air supply is forced through the powder it may rise by as much as 50 mm. Switch on the forced air supply and notice that the air passing through the powder makes it appear as if it were a boiling fluid. Next, suspend the heated article in the fluidising powder for about 20 seconds. Remove the article, tap its suspending wire to knock off any surplus powder and hang the article on a nail or hook to cool. Then pick-up some of the powder with the heated end of a strip of wire, and work the resulting molten polythene into the part of the metal which was previously covered by its suspending wire.

Lathes and turning

Lathes are for working mainly cylindrical, conical and curved forms in materials as they are rotated against hand-held or machine-held tools. The process of forming materials to shape in a lathe is known as *turning*. The two types of lathe common in schools are usually powered by electric motors. They are *wood-turning lathes*, on which wood can be turned with hand-held tools, and *centre lathes*, on which metals and plastics can be turned with tools held in tool posts fitted to the lathe.

Wood turning

Wood-turning lathes

A typical motor-powered wood lathe consists of a *cabinet pedestal* in which an electric motor generating between 500 and 1000 W, is mounted. A four-step pulley mounted on the motor spindle is connected by an endless V-belt to a headstock in which a spindle with a similar four-step pulley is mounted between ball and roller bearings. By positioning the endless belt between the different size pulleys on motor and headstock spindles, a range of speeds can be obtained to drive the wood as it is turned. Typical speeds range from about 400 revolutions per minute (rev/min) to about 2700 rev/min. The lower speeds are for turning wood of large diameter, the higher speeds for turning wood of small diameter.

The headstock spindle has screw threads cut on each end – a right-hand screw on the right-hand end, a left-hand screw on the left-hand end. Chucks and faceplates can be fitted on the screwed ends (these fittings are described below). A hole is bored right through the headstock spindle and each end of this hole is further taper bored to receive fitments which have Morse taper spindles. The centres of a wood lathe would typically be at about 150 mm above the surface of the lathe bed. A tailstock and a tool rest are fitted to the lathe bed in such a manner that they can be locked in any position along the length of the bed.

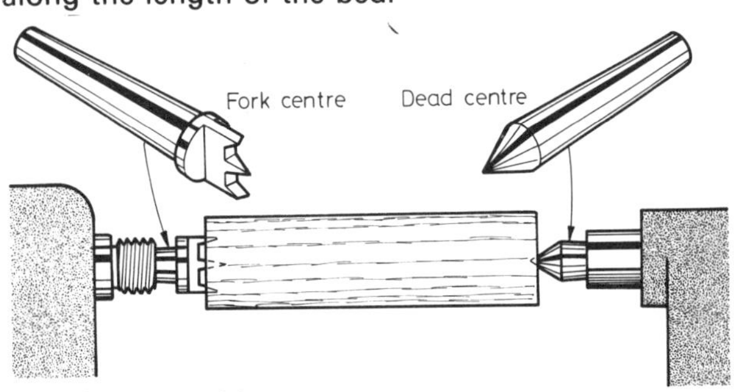

Wood mounted between centres

Work up to nearly 300 mm in diameter can be turned over the bed of a 150 mm centre lathe. Larger work, up to about 500 mm in diameter, can be turned on wood fitted to a faceplate screwed on the left-hand end of the spindle.

Turning between centres

This method of holding wood in a lathe is for turning long pieces.

Fit a fork centre in the tapered hole of the headstock spindle and saw a slot in one end of the wood to fit over the fork centre. Lock the tailstock in a convenient position along the lathe bed and fit a dead centre in the tapered hole of the tailstock spindle. Force the dead centre into the end of the wood with pressure exerted by turning the tailstock handwheel – the fork centre in the headstock will force the wood to rotate when the lathe motor is switched on. The other end of the wood rotates on the dead centre.

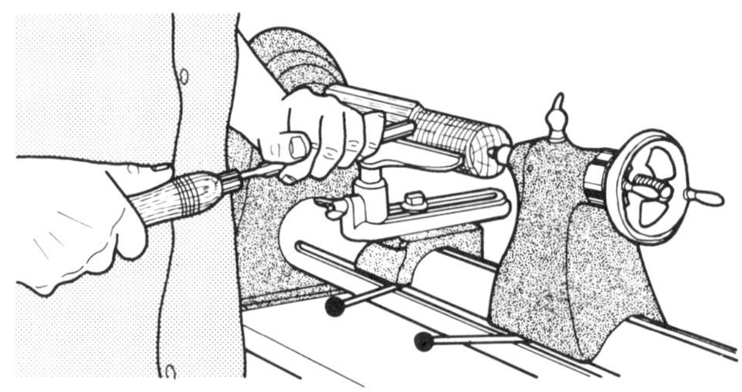

Turning wood with a wood-turning gouge

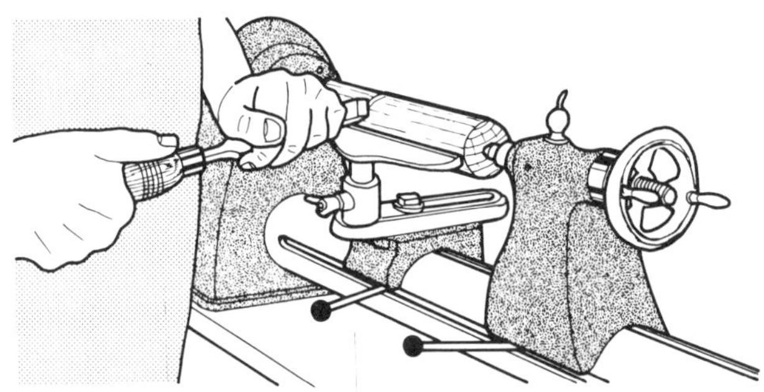

Turning wood with a wood-turning skew chisel

Two drawings show the process of turning wood with a gouge and a turning chisel and illustrate the stance of the person as the two tools are handled. A fuller description of hand wood-turning tools is given on page 25. To prevent the wood burning as it rotates on the dead centre, add a spot of oil, grease or tallow.

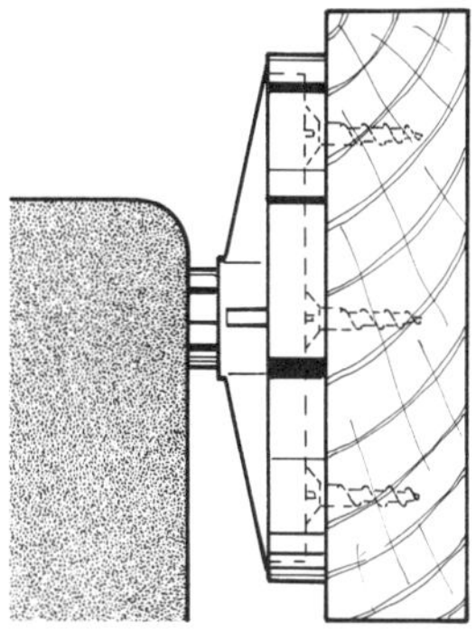

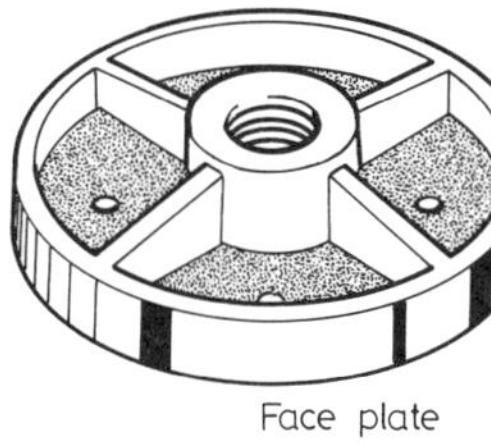

Wood mounted on a faceplate

Turning wood on a faceplate

This method is for turning wood of large diameter.

A piece of wood, sawn roughly circular, is fixed with wood screws on a faceplate which is itself screwed to the end of the headstock spindle. The lower spindle speeds should be selected when turning wood of large diameter. Although turning gouges and chisels can be used when turning large diameter work, their manipulation requires considerable skill. In school work much of the turning of large diameter wood will be carried out with scraper chisels which require less practice and skill to manipulate.

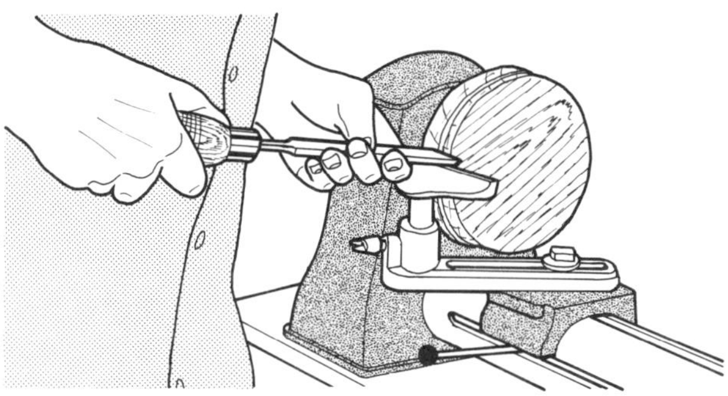

Turning with a wood-turning scraper chisel

Turning metals and plastics

Centre lathes

Centre lathes are for turning metals and plastics. Their basic construction and mode of operation are similar to lathes for wood turning. A major difference is that the tools employed for cutting the materials being turned in a centre lathe are fixed firmly in a *tool-post* mounted on a compound slide which is, in turn, mounted on a carriage. The carriage can be moved along the lathe bed under the control of a hand wheel. The compound slide is equipped with controls for moving the tool held in the tool-post along, across, or at an angle to, the length of the work as it is being turned. Another difference is that centre lathes are fitted with a more complex system for changing spindle speeds, which allows the much lower spindle speeds necessary when turning hard metals of large diameter to be achieved.

Lathe fittings

The principle fittings for holding work for turning in a centre lathe are ***chucks, faceplates*** and ***centres***. A *three-jaw self-centreing* chuck will grip rods or material of hexagonal section centrally, as the three jaws are automatically closed together by operating a square-ended key in any one of the three square slots spaced on the outside of the chuck. For mounting work which is not cylindrical in section *four-jaw independent* chucks are suitable in which each jaw is operated independently with a square-ended key.

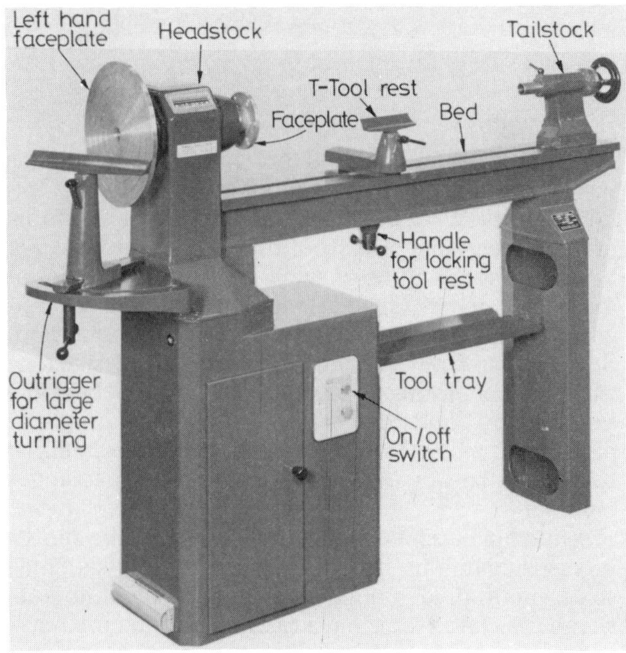

A 150 mm centres wood-turning lathe (Photo by courtesy of Denford Machine Tools Ltd.)

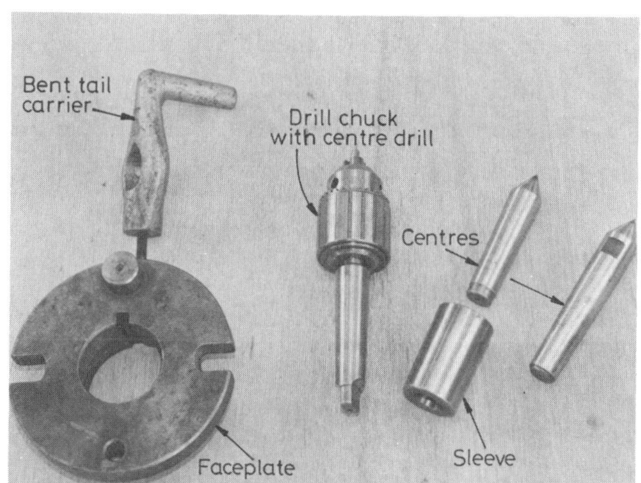

Lathe accessories for holding work to be turned between centres

Faceplates and centres are similar to those for wood lathes, except that those for centre lathes are of heavier construction to withstand the greater strains imposed on the lathe when metals are turned as compared to the strains involved in turning the softer material, wood.

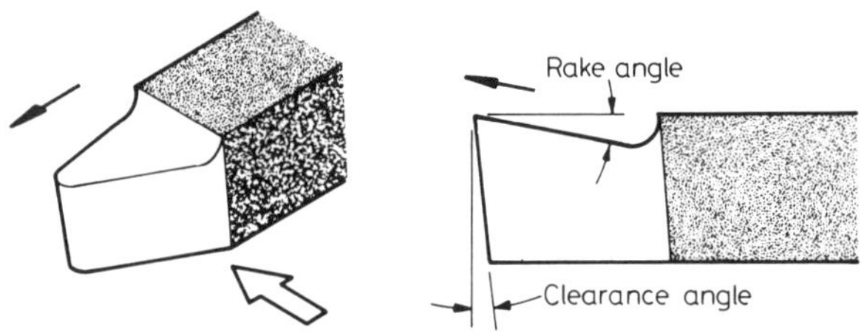

Rake and clearance angles of a lathe tool

Lathe tools

Lathe tools for fitting in the tool posts of centre lathes will either be small *tool bits* which are held in *tool holders* or they will be individual solid, one-piece tools of larger sectional size. Tool bits vary in size but are basically short lengths of small, square, rectangular or round section high speed steel. Tool bits are discarded when they become small after repeated re-grinding. Solid, one-piece lathe tools are made from steels of larger sectional sizes. Although lathe tools can be made from carbon tool steel, they are more often made of high speed steels, which will cut at higher rotational speeds and at higher temperatures without becoming blunted.

A number of different shapes of lathe tool are shown in the illustrations here, but these are only a small selection from a much larger variety of lathe tool shapes. All lathe tools have two features in common – the angle of *rake* and the angle of *clearance* (shown below). In the illustrations, small arrows placed near to each plan drawing indicate the direction of the rake angle of the tool (the arrow is pointing up the rake angle). As a lathe tool becomes blunted after prolonged use, the tool must be carefully reground, maintaining rake and clearance angles while doing so. Rake and clearance angles should be varied according to the material being turned. Some examples are given

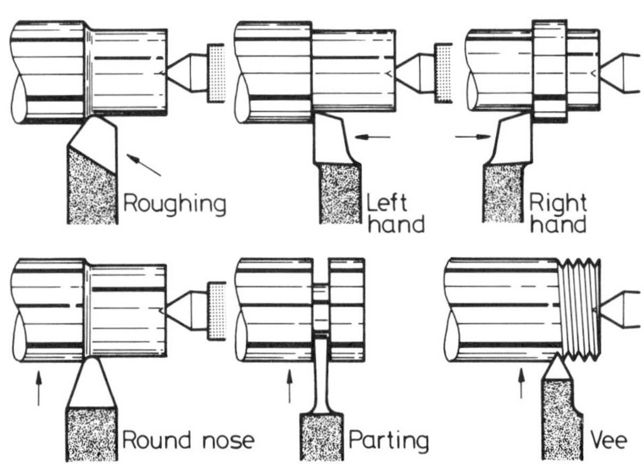

Lathe tools – arrows show direction of rake angle

in Table 1. The speed at which various materials should be cut should also be varied with the material. From the speeds in metres per minute given in Table 1, the rotating speed to which a lathe should be set can be calculated from the formula:

$$\frac{\text{Speed in metres per minute} \times 1000}{\pi \times \text{diameter of material}}$$

Table 1

material	rake angle	clearance angle	cutting speed (metres per minute)
cast iron	0°	5°	25
brass	0°	10°	90
mild steel	20°	8°	25
aluminium	30°	10°	300
nylon	30°	10°	250

All lathe tools should be set in the tool post of the lathe so that the working tip of the tool is exactly in line with the lathe's centres (see photograph). The height

Fitting one end of a rod to the lathe

Setting cutting tool to dead centre

of the tool can be varied in the tool post by adding, or taking away, a variety of different thicknesses of steel packing strips placed below the tool in the tool holder.

Turning between centres

When material is held in a three-jaw self-centreing chuck for turning, no matter how accurately made is the chuck, or how precisely cylindrical is the outside of the piece being held, once the material is removed from the chuck it is very difficult to replace it exactly central again. Thus, precision turning which involves *re-chucking* (replacing the work within the chuck) presents difficulties when a three-jaw self-centreing chuck is used. When turning material between centres, however, the work can be taken from the lathe centres as many times as necessary to complete a required shape without any loss of correct re-positioning between the lathe centres.

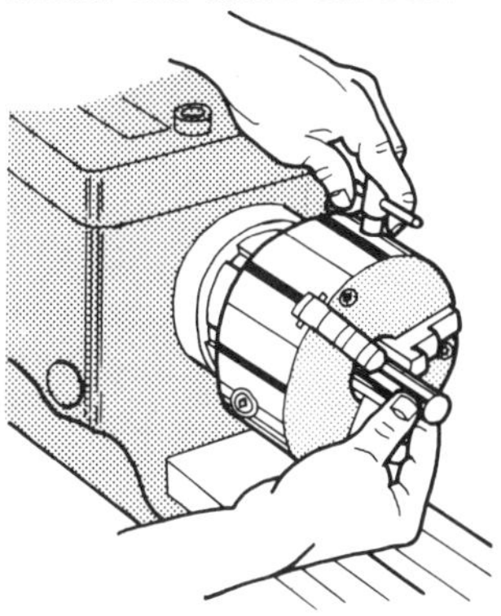

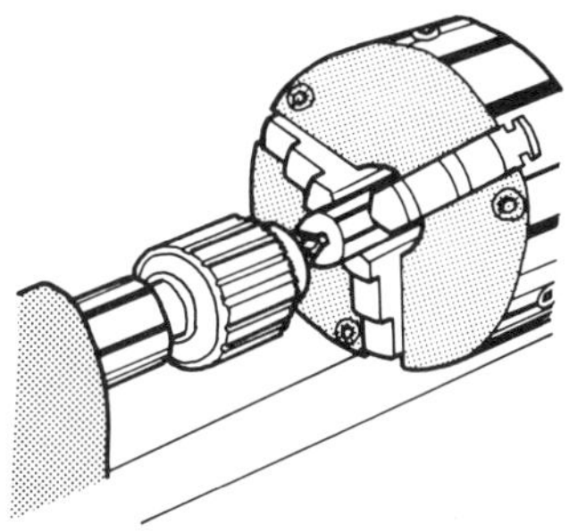

Drilling in the lathe with a centre drill

Fitting a rod into a self-centreing three-jaw chuck

A series of drawings (above) and photographs (opposite) illustrates the stages involved in turning between centres.

1. Place the rod which is to be turned into a three-jaw self-centreing chuck. If the piece is long, the hole through the headstock spindle will allow the rod to be held as shown in the drawing. Tighten the chuck jaws with the chuck key.
2. Fit a centre drill into a drill chuck and fit the drill chuck into the tailstock. Under the pressure exerted by turning the tailstock handwheel, bore a centre hole in the rod end. Then release the rod from the three-jaw chuck, turn it around and bore a centre hole in the other end.
3. Screw a *bent-tail carrier* on to one end of the rod and fit a catch faceplate on the headstock spindle and a centre in the spindle's tapered hole. Then place the centre-drilled hole in the end of the rod on to the headstock's centre.
4. Place the other centre-drilled end of the rod on a centre in the tailstock spindle and lock the tailstock in position on the lathe bed. Turn the tailstock handwheel to tighten the rod between the two centres and tighten the tailstock spindle in the tailstock. Place several drops of oil, or a smear of grease, on the tailstock centre to assist the easy running of the rod on the centre.
5. The work can now continue by fitting appropriate lathe tools in the lathe's tool post and, with the various hand controls on the saddle and compound tool rest, the required shape can be turned.

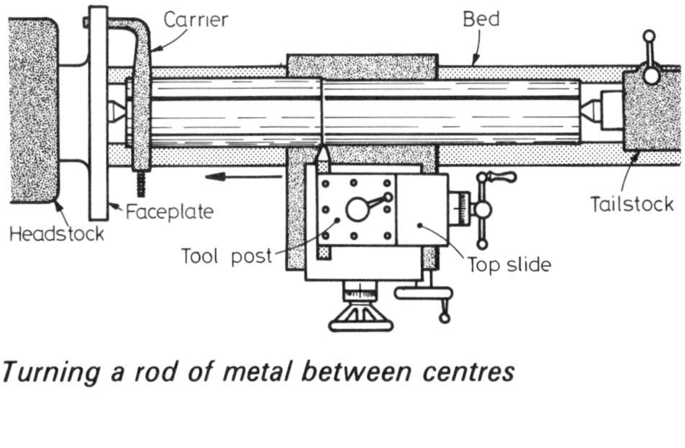

Turning a rod of metal between centres

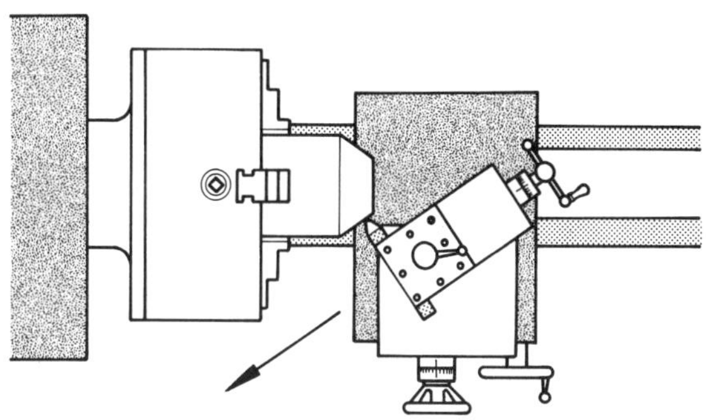

Taper turning in a chuck

Turning tapers

Tapers of short length can be turned by placing the work in a three-jaw self-centreing chuck and then positioning the top slide of the compound slide rest at the angle at which the taper is to be cut. The tool in the tool post is then made to make a series of cuts along the slope of the taper under the rotation of the top slide hand control.

Long tapers are best turned between centres. The tailstock of a centre lathe can be moved forwards or backwards at right-angles to the lathe bed and then locked in a chosen position. To work a taper on a long rod held between the lathe centres, the tailstock is moved forwards or backwards to the position required by the taper, and normal cuts are made along the rod parallel to the lathe bed. If the tailstock has been moved forwards, the right-hand end of the rod will become thinner as the cutting proceeds; if the tailstock has been moved backwards, the left-hand end of the rod will become thinner. In this way a taper will be formed.

Machines

Six photographs on this page show other machines which will be seen in school and college workshops. Note that some of the machines in schools are not for the use of pupils, but are intended only for use by instructors, teachers and workshop technicians. The bandsaw shown here is one such machine.

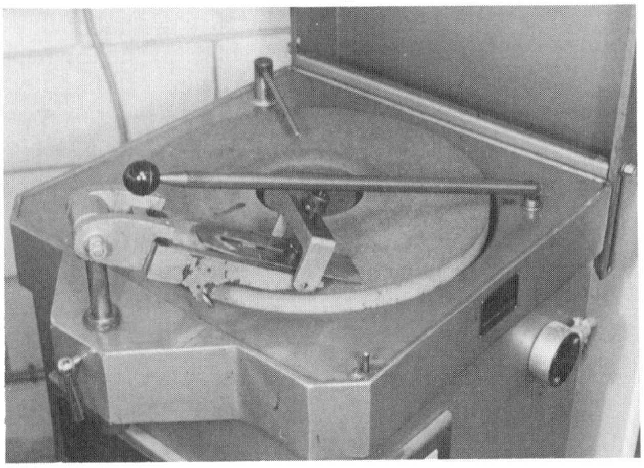

Denford tool grinding machine

Carborundum wheel grinder

Changing speeds of a drilling machine

Sanding machine

Sawing with a bandsaw

Bench drilling machine

Heat treatment of metals

Steels

The steels for hand craft work are mostly *carbon steels* – essentially alloys of carbon and steel, although other materials may be included, e.g. manganese (up to 1.5%) or silicon (up to 0.5%). The carbon content determines the character and thus the uses of a carbon steel. Four common carbon steels are listed here.

Dead mild steel – carbon content up to 0.1 %. This steel is soft and malleable and used for soft iron wires.

Mild steel – carbon content 0.1 % to 0.3%. This is the most common carbon steel. It is tough, machines well and is often used in constructional craft work.

Medium mild steel – carbon content 0.3% to 0.6%. This is a tough and hard metal from which high tensile wires and high tensile bolts are made.

Carbon tool steel – carbon content 0.6% to 1 %. This is a tough and hard steel which can be hardened and tempered to take a cutting edge. It is the steel from which many of the tools in school workshops are made.

Hardening and tempering of carbon tool steel

Only carbon tool steels with their 0.6% to 1 % carbon content can be hardened and tempered. The processes involved in workshop hardening and tempering are as follows.

1. Polish the surfaces of the piece being hardened and tempered with emery cloth.

2. Heat the piece until it glows cherry red, then quench it in water. This sudden lowering of the temperature of the heated metal renders it very hard.

Hardened tool steel is not of much practical value for if it is subjected to a sharp blow it will fracture. A test to check whether the steel has become hard is to attempt to file it with an old file. The file should not cut the hardened steel, because it will have become harder than the material from which the file is made.

3. Polish the now blackened surfaces of the hardened steel with emery cloth.

4. Carefully re-heat the hardened steel. If only the cutting edge of the work is to be tempered, for example, as when making a cold chisel or a centre punch, heat can be applied with a blowlamp at a point about 50 mm behind the cutting edge. The heat will be conducted along the metal to the cutting edge. For tools requiring an overall tempering, very careful overall heating is necessary. As the temperature of the metal increases from cold, the polished surfaces will become coloured as oxides form on the surfaces. The first oxide colour to appear will be straw, which changes to brown, then purple, followed by blue and, finally, black.

When the oxide colour indicates that the hardened steel has reached the desired tempering temperature, the piece is again quenched to cold by dipping in water. As the temperature increases so the brittleness of the hardened steel is tempered until, when the oxides become black, the steel is again too soft to take a cutting edge. The following is a guide to the oxide colours which indicate particular tempering temperatures.

Straw – lathe tools for cutting metals.
Brown – plane cutters, wood chisels.
Purple – wood saws, cold chisels, knives.
Blue – screwdrivers, springs.

Case hardening of mild steel

A very thin layer of hardened steel can be produced as a skin to mild steels. The metal is heated to red heat, then dipped into a proprietory *case hardening powder*. The metal is then re-heated to cherry red and quenched in water to produce a thin hardened skin or *case* on the mild steel. Proprietory case hardening powders used in this manner add carbon to the skin of the mild steel. The depth of the hardened skin is probably about 0.05 mm.

Annealing

Mention has already been made of annealing metals such as copper. See page 38.

Annealing is a process employed when metals require to be softened, either because the metal has been hammered or otherwise worked and become *work hardened*, or because the metal is too hard to be satisfactorily worked with tools and must be softened before work can commence. Annealing involves heating the metal and then either quenching it to a cold state in water, or allowing it to cool slowly.

Methods of annealing specific metals are described below.

1. Carbon steels are annealed by heating to cherry red and then allowing them to cool very slowly. In a workshop this can be done by heating the steel in a forge and then allowing the forge to cool down while the metal is still within the forge.

2. Copper is annealed by heating to cherry red and then quenching in water.

3. Brass is annealed by heating to cherry red and then allowing it to cool slowly in air.

4. A method of annealing aluminium is to coat it with a layer of soap, heat the metal until the soap turns black, and then to allow the metal to cool. The temperature of burning soap is that at which aluminium softens after becoming work hardened.

Annealing is sometimes referred to as *normalising*.

Maintenance of tools and equipment

Good results cannot be achieved in craft and technology unless tools and their associated equipment are maintained in a clean and sharp condition. A knowledge of methods of maintaining tools is an essential feature of craft skills.

Benches and bench equipment

1. Bench tops should be clean and kept as smooth as possible.
2. Make good use of sawing and chiselling boards and chipping blocks to avoid damage to bench tops.
3. Re-rack tools when you have finished working with them.
4. Place polythene sheet on a bench top when working with plastic resins such as polyester resin.
5. Avoid placing hot metal on a bench top.
6. Oil, grease, and spots of glue on a bench top can spoil your work.
7. Do not hammer down on the slide of an engineer's vice.
8. The screws of vices and cramps should occasionally be lightly oiled.

Marking and measuring tools

1. Try squares, straight edges, rules, micrometers and other marking tools are precision tools and should be treated as such.
2. Knives, punches and the spurs and cutting blades of gauges will need sharpening from time to time.

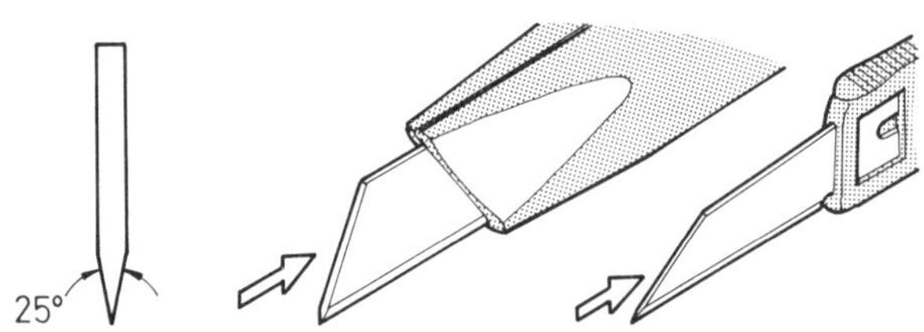

Sharpening angles for marking or cutting knives

Saws

1. Wood saws will not cut metals. Avoid contact between the teeth of a woodsaw and other tools, and never saw through a nail with a wood saw.
2. After prolonged use most wood saws will require sharpening and resetting for which a saw file and setting tools are required. Saw sharpening and setting are skilled tasks which should be left to teachers, instructors and workshop technicians.
3. Disposable saw blades are expensive, e.g. hack saw, coping saw, piercing-saw blades. They will last a long time if used with care. Avoid over-tightening, buckling or incorrect sawing procedures.
4. Wood saws will slide in their cuts more easily if candle grease is smeared on each side occasionally.

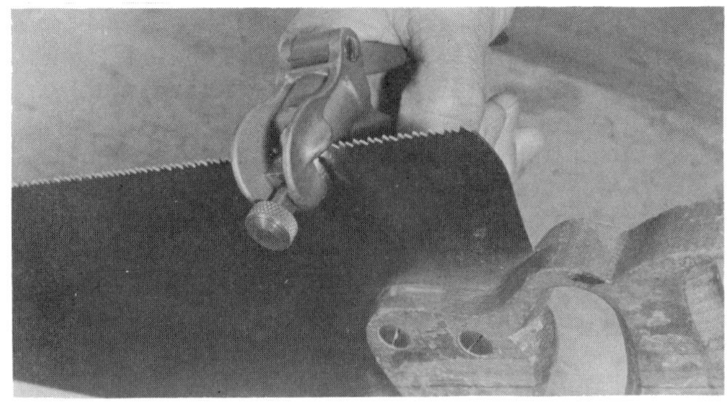

Setting saw teeth with saw setting pliers.

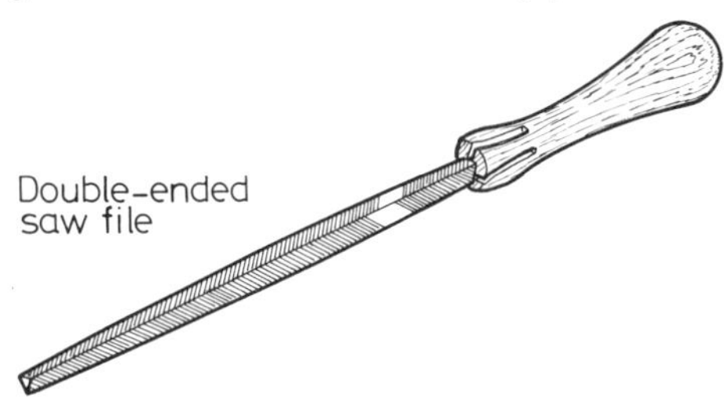

A saw file

Planes and files

1. Plane blades need to be regularly sharpened, and when the width of the sharpening angle becomes too great, they must be re-ground. A Denford tool grinding machine designed for this purpose is shown on page 56.

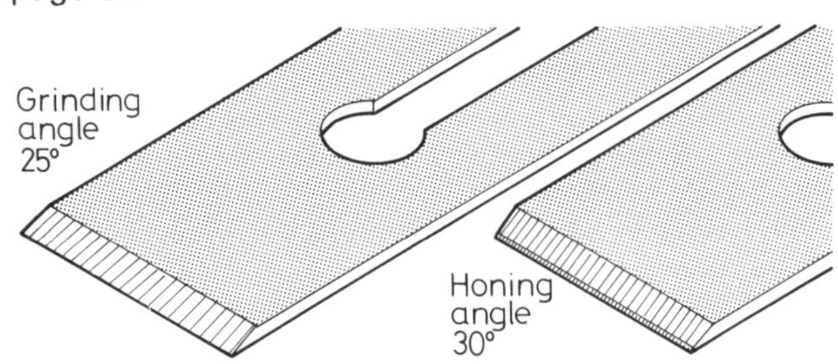

Grinding and sharpening angles for a plane cutter

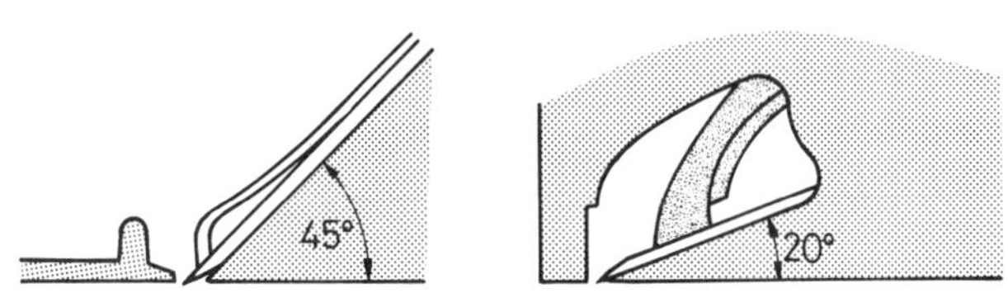

Angles at which cutters are placed in plane bodies

2. A plane cutter is sharpened as follows.
(a) Remove the cutter from the plane body and remove the cap iron from the cutter.
(b) Place drops of thin lubricating oil on the oilstone surface.
(c) Hold the cutter at 30° to the oilstone surface and work it over the surface under hand pressure using, as far as possible, the whole surface of the stone.

Sharpening a plane cutter on an oilstone

(d) When a burr forms on the back of the cutting edge, turn the cutter over and rub it over the stone holding it quite flat on the stone's surface.
(e) Repeat (c) and (d) several times to loosen the burrs.
(f) Push the sharp edge into a piece of softwood to remove the burr.

3. Files should be placed in racks or purpose-made trays when not in use so as to prevent files rubbing against each other and so damaging the cutting edges.
4. Keep the teeth of a file open with the aid of a file card which will brush waste from between the teeth of a file.

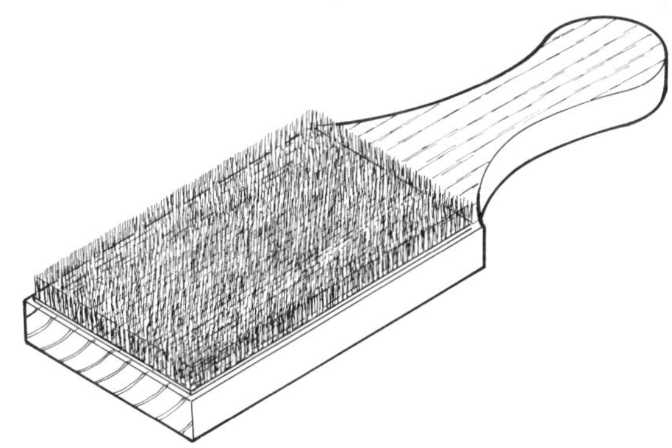

A file card

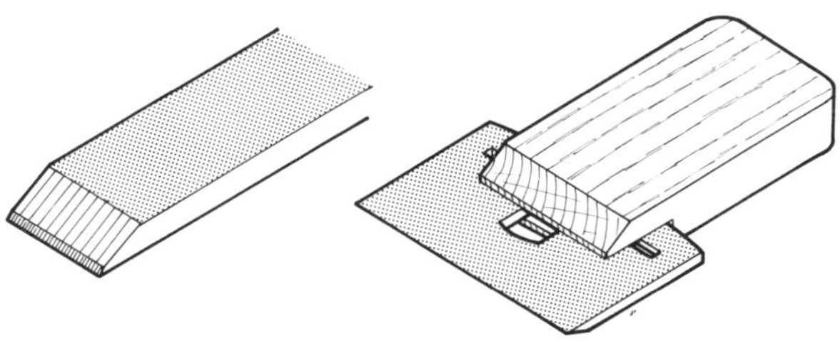

Sharpened edge of chisel (left) and a wooden spokeshave cutter holder (right)

Spokeshaves
Spokeshave cutters are sharpened in a similar manner to plane cutters. The problem of handling such short cutters can be overcome by using a workshop-made spokeshave cutter holder.

Chisels and gouges
1. Wood chisels are sharpened using the same routine as when sharpening plane cutters.
2. The cutting edges of cold chisels are ground on a machine grinder to preserve the cutting angle of 60°.
3. Gouges are sharpened with the aid of *oilslips* – cylindrical or semi-cylindrical pieces of oilstone. The oilslip is used to sharpen the inside curved edges of gouges, and a flat oilstone is used for the outside curved edges.
4. Carving gouges are easier to use if a slight angle of 5° is sharpened on the inner cutting face.

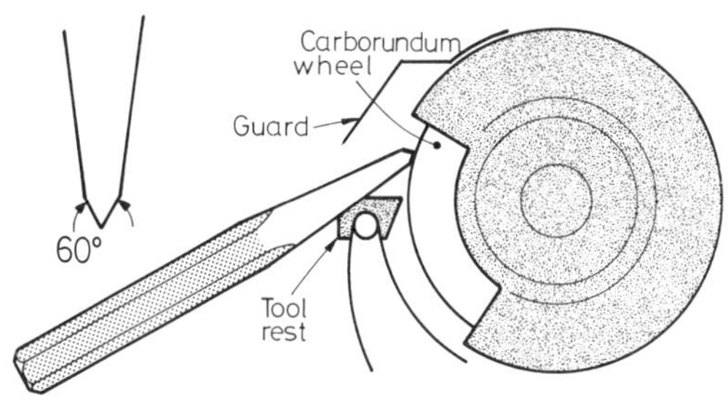

Grinding a cold chisel

Hammers
1. Hammers should be kept in purpose-made racks.
2. The peins of hammers should be kept clean and polished. This can be achieved with the aid of fine emery cloth or with fine sandpaper.
3. Hammer heads must be held securely on their handles so check the handle wedge occasionally. If a hammer head works loose it should be re-wedged.

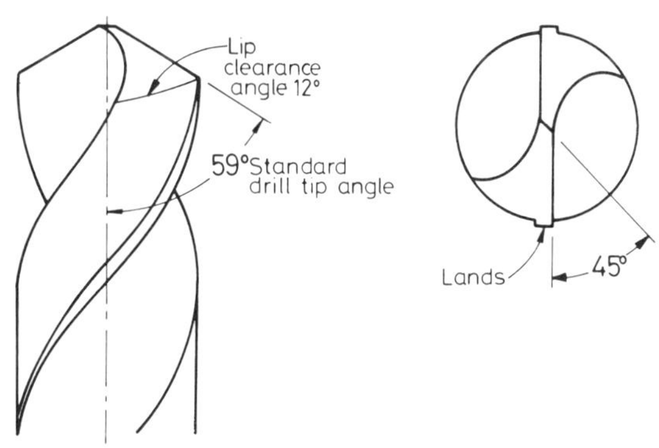

Grinding angles of a twist drill

Drills and bits
1. Twist drills must be re-sharpened when blunt. This is preferably carried out on a machine grinding wheel. Drill grinding jigs are made which, in unskilled hands are advisable.

2. The cutting angle, clearance angle and angle of the drill tip must be accurately retained when drills are sharpened. If the angles are incorrect the drill will not bore accurate holes.
3. When wood bits require re-sharpening, saw files and needle files can be used. Care must be taken to avoid reducing the width of the bit.

Screwdrivers

1. Pozidriv and Phillips tipped screwdrivers cannot be re-sharpened but the tips are hard and tough.
2. Straight-tipped screwdrivers can be re-sharpened on a grinding wheel. Check that the tip fits the screw slot for which the screwdriver is intended.

Taps and dies

Taps and dies, for forming screw threads, are made from hard steels. They will break relatively easily if used carelessly so never force these tools to cut more quickly, or to make deeper cuts, than they are capable of. Keep the work and tools lubricated with a suitable lubricant.

Forging tools

1. Keep forging tools in their proper racks.
2. The area around a forge and anvil should be kept as clear as possible.

Soldering irons

1. Soldering irons must be kept tinned (see page 41), ready for use.
2. Check the cables and plugs of electric soldering irons occasionally to determine that the cables are not frayed or damaged and that the plugs are secure on the cables.

Working with polyester resins

1. Place polythene sheet on benches on which the work is to take place.
2. Clean all brushes, rollers, knives and other tools immediately after they have been used.
3. Untidy and disorganised methods of working can result in resins setting at the wrong times and in the wrong places. This can result in spoiled work.

Turning tools

1. Lathes are expensive machines and they should be treated as such.
2. Do not commence turning until you are sure that the lathe will be rotating at a correct turning speed for the material being worked.
3. After use the machine and the area around it should be cleaned. All wood waste or waste metal turnings (*swarf*) should be removed.
4. Remove waste wood or metal from around the machine before it builds up in sufficient quantities to become a nuisance to the person operating the lathe.
5. Wood hand-turning tools are sharpened in a similar manner to that in which wood planes, chisels and gouges are sharpened, except that turning chisels are sharpened from *both* sides of the blade and scraper chisels have a sharpening angle of between 60° and 70°.
6. When re-grinding lathe tools intended for metals or plastics, be careful to retain the correct rake and clearance angles.
7. Use a thread tool gauge to check the angle of a V tool for cutting screw-threads when re-grinding.

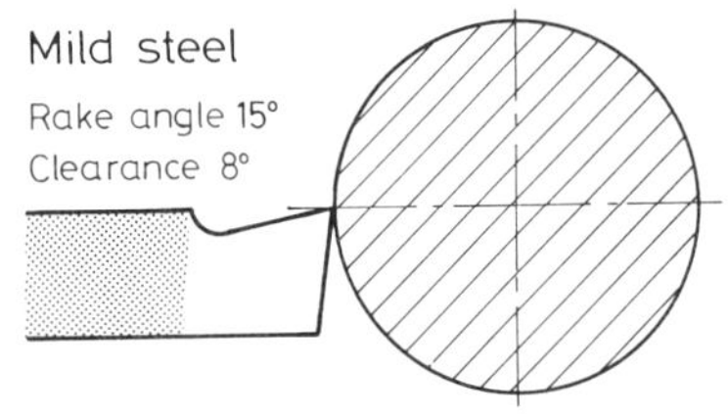

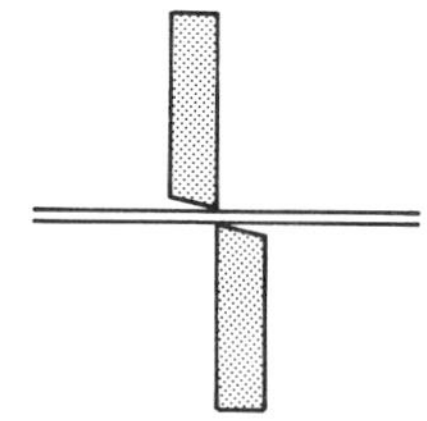

Sharpening angle of tin shears

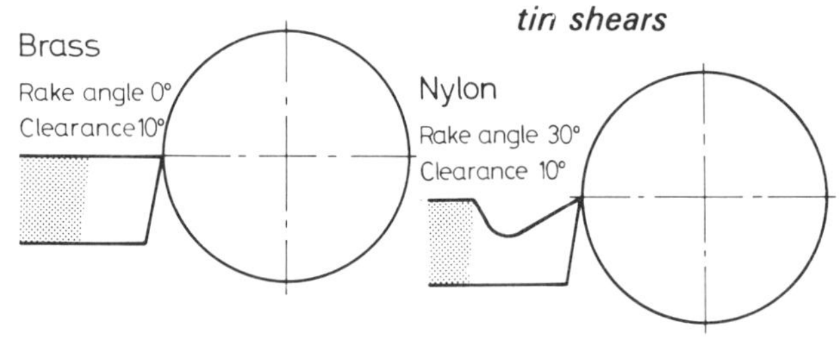

Rake and clearance angles of lathe tools

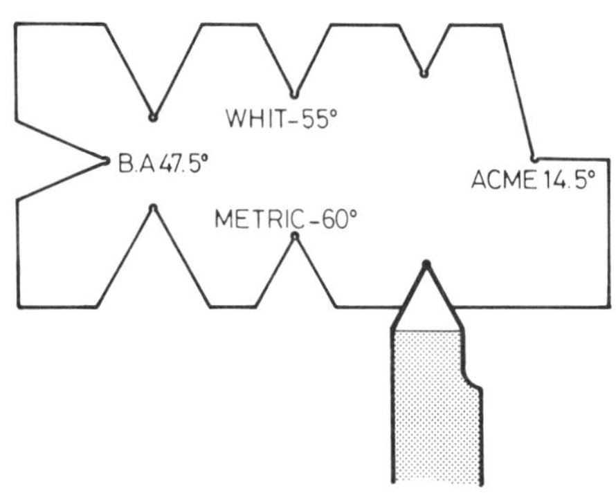

A screw-thread gauge

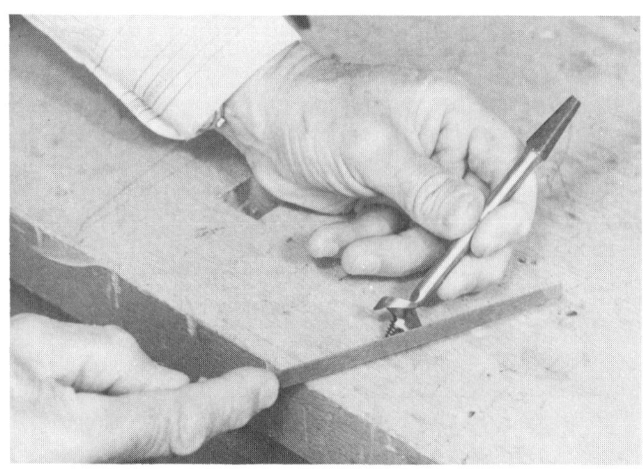

Sharpening a centre bit with a saw file

Safety

Details concerning the safety precautions which should be observed by all those engaged in craft and technology are dealt with on this page and on page 62 overleaf. Despite the information given in these two pages, a first rule for the adoption of safe working practices in craft rooms is that safety is often a matter of common sense. All people working in craft rooms should take those precautions which any thinking, sensible person should take to ensure the safety not only of him or her self but of those around.

Dress

1. Wear an apron or an overall. Not only does this assist in keeping clothing clean, but it also holds loose clothing in place.
2. Although ties can be tucked under aprons or overalls, they are best not worn at all.
3. When working at a forge or when casting molten metal, wear heavy chrome-leather aprons and leggings which give over shoe protection. In addition, when casting molten metal, wear chrome-leather gloves. The chrome leather will protect body and hands from being burnt by small pieces of molten or hot metal. A visor, to cover the face, must also be worn.
4. Hard shoes or boots should be worn. Soft shoes such as plimsolls or training shoes are not suitable. Heavy items of equipment, tools or pieces of metal falling from, say a bench, may cause injury to feet not properly protected.
5. Loose jewellery such as bangles and necklaces are best not worn at all. They may be the cause of an accident by being caught by a tool or in the moving parts of a machine.
6. Long hair can present a safety hazard when working near a machine or near a lighted blowlamp, and it is best worn under a net or under a suitable cap.
7. Shirt sleeves should be rolled, even when covered by the arms of an overall.
8. When working at some machines – grinders, sanders or lathes – it is advisable to wear goggles with clear transparent anti-fracture lenses.
9. When welding, protective goggles with the correct eye lenses *must* be worn. Serious eye injury can result from not wearing such protective goggles.

Movement

1. Do not run when moving around a working area as fast movements are distracting, and may cause accidents.
2. Do not fool about; play can be dangerous. Workshops are not places for practical jokes.
3. Carry tools in a way which will not be hazardous to yourself or to other people. Carry sharp-edged tools 'sharp edge down'.
4. Make certain when carrying a piece of material that you know where both its ends are. A long swinging metal rod carelessly carried can be very dangerous.

Act sensibly

1. Never work alone in a workshop, in case something goes wrong and you need urgent assistance.
2. There will be a choice of equipment for any piece of work. School and college workshops will be well stocked with the tools and machines you need. By using an inappropriate machine you may cause injury, or risk damaging your work.

Cleanliness

1. Keep yourself clean. Wash your hands after working with materials which soil them. Some materials (e.g. resins, glues) are skin irritants; others may prevent a tool from being held securely.
2. Do not wear dirty aprons or overalls.
3. Do not allow tools or pieces of material to accumulate on a bench or near the area in which you are working. They may get in your way and cause you to slip or to fall, or tools may be knocked from a bench because of overcrowding. A falling sharp chisel can cause severe cuts in legs and feet.
4. If tools have become dirty, as for example, if used to remove glue, or when brushes have been used to lay-up polyester resin, see that the tools are cleaned after use. A sharp-edged tool soiled with hard glue or with grease or oil can be potentially dangerous.

Tools

1. Tools must be kept well sharpened. They do not cut efficiently if blunt or wrongly sharpened, and may slip under undue force to make them cut, causing an accident.
2. To operate correctly, tools must be set properly.
3. To prevent hand injuries with sharp-edged tools, a good rule to follow is 'both hands on the tool'.
4. Tools should be properly racked to avoid injury from one tool when reaching for another. Sharp-edged tools should be racked with the edges guarded, and heavy tools must be securely racked to prevent falling.
5. Tools like micrometers and machine gauges should be kept and placed on racks in the wooden boxes they were sold in.

Machines

Before switching a machine ON

1. Before preparing any machine, or adjusting it before commencing work, make sure that the mains switch connecting the machine to the power supply is in its OFF position.
2. Check that all guards are in position on the machine – such as guards covering moving pulleys and rotating gear systems.
3. Any work held in chucks or other holding devices must be firmly and securely held. If a chuck key has been used to tighten the work in a chuck, remove the key before the machine is switched ON.
4. Make sure that tools in the machine are held securely in place, e.g. lathe tools in a tool post, drills in the chuck of a drilling machine.
5. Remember that some metals cut more easily when a coolant is run between the work and the tool. Proper application of coolants assists in preventing those accidents which can arise from work breaking or buckling in a machine.
6. Before switching ON check that the machine is running at the speed suitable for the material and the size of the material being worked. If necessary adjust the machine to run at the correct speed.

After switching the machine ON

1. Do not attempt to remove swarf with your bare hands as it is cut from the machine.
2. Do not touch moving parts of the machine.
3. Make no attempt at any time to slow down or to stop a machine as it is rotating by pressure of the bare hands. This applies even if the machine has been switched OFF and is beginning to slow down before stopping.
4. Do not attempt forcing a tool to cut deeper or faster than it should. Either the work will be damaged or the tool will be broken. Both may be a source of danger.
5. When drilling with a drilling machine, the work being bored must be securely held in a machine vice or in hand-held hand vices, grips or clamps. Never hold such work in the fingers.

Hand machine tools

Some powered tools are hand-held, examples being hand power drills and sanders. Hand-held powered tools can be a source of considerable danger unless properly and safely used. However if properly used for work within their capacity, they can be quite safe and they will save considerable labour. The same rules of safety should be followed when using hand-held powered tools as when using other machines. Also:

1. Hold the machine in both hands. Then there can be no injury to the operator's hands.
2. The cable connecting the tool to a power supply must be in a safe position – not in a position in which it may trip people, nor in a position where it could be cut by the tool itself. A good rule is to keep the cable as short as is possible. Another good rule is to loop the cable over one's shoulder.
3. Hold the piece being worked securely in a vice, a cramp or in some other holding device.
4. Do not apply unnecessary force. Too much pressure may break tools held in the power unit possibly resulting in danger. In any case, too great a pressure may result in inefficient working of the machine.
5. Do not attempt to use drills or cutters in hand power tools which are larger than the capacity for which the tool is designed.
6. When changing drills, bits or other cutters, switch OFF at the mains power supply and then pull out the plug connecting the cable to the power supply. Only then should you attempt to change tools in the machine.

Working with plastics

Polyester resins

When working with polyester resins and glass fibre the following precautions should be observed:

1. Always use a barrier cream on the hands. It is also advisable to wear disposable polythene gloves.
2. If glass fibre irritates the hands, apply barrier cream and wear disposable polythene gloves or rubber gloves.
3. If catalyst gets into the eyes, wash the eyes thoroughly with plenty of water. Then consult a doctor. Catalysts can irritate and burn the skin. Wash them off immediately with plenty of water.
4. Waste catalysed resins can generate enough heat to start a fire. Place such waste in metal containers or dowse the waste in cold water. Do not place it among rags or papers.
5. Cleaners used for cleaning off uncured resin are very inflammable. Be careful with rags impregnated with such cleaning fluids and do not use these cleaners to clean the hands. Wash the hands with soap and water or use special hand-cleansing creams.
6. Good ventilation of the working area is essential.

Using other plastic materials

1. When cutting expanded polystyrene with a hot wire cutter, the styrene fumes emitted can irritate the eyes and even cause drowsiness. Allow adequate ventilation. Do not allow the cutting wire to run red hot.
2. When filing and sanding rigid polyurethane foam to shape, wear a breathing mask with suitable filters. The fine dusts arising from rigid polyurethane foam can stick to the lung linings and cause irritation.
3. After applying any of the resin adhesives, wash the hands. These adhesives can cause skin irritation.

Other safety precautions

When sawing some materials irritant dusts are given off which can cause quite severe discomfort. Thus, when sawing, filing or sanding materials such as chipboards or GRP mouldings it is advisable to wear an appropriate breathing mask.

Exercises

1. What protective clothing should be worn when:
(a) working at a wood bench;
(b) welding with oxy-acetylene equipment;
(c) casting aluminium in a sand mould;
(d) grinding a cold chisel at a carborundum wheel;
(e) working with polyester resins.
Why is it necessary to wear such protective clothing?
2. Name three types of screwdriver. What are the advantages of each type of screw for which the three screwdrivers are made?
3. Explain, with the aid of sketches, the methods of screwing two pieces of wood together with countersunk wood screws.
4. Sketch four forms of rivet. Name materials from which the rivets you have sketched would be made. Explain in detail how you would make a joint involving countersunk iron rivets.
5. What is meant by the terms *tap* and *die*?
6. Explain, with the aid of sketches, how you would cut a screw size M8 on a length of mild steel rod of diameter 8 mm.
7. Why is it important to use a lubricant when cutting threads in some materials? Name materials which do not require a lubricant when cutting screw threads.
8. What is meant by the term *vacuum forming* when applied to the working of plastic materials? Give examples of items which you consider can be vacuum formed.
9. Explain in detail, with the aid of sketches, how you would form a right-angled bend along the length of a sheet of plastic of 2 mm thickness and 150 mm length.
10. What are meant by the terms raising, sinking, blocking, planishing? Use sketches to illustrate your answers.
11. Make a sketch of an anvil, and name its various parts.
12. Explain how you would form an eye of diameter 100 mm at the end of a mild steel rod, itself of 10 mm diameter.
13. Why are soldering irons *tinned*? How would you tin a soldering iron which you had just purchased?
14. What is meant by the term *heat sink* as applied to soldering? Give an example of heat sink soldering, explaining why you have chosen the example you gave.
15. Explain the differences between brazing spelter and silver solder. Give examples of work which should be brazed and further examples of work which should be silver soldered.
16. Name a resin which you can use for encapsulation of a biological specimen. What is the purpose of such encapsulation? How can you ensure that the surfaces of the encapsulation are quite smooth?
17. Explain in detail, with the aid of sketches, the processes of forming a glass reinforced plastic moulding.
18. What are the advantages of GRP mouldings over similar shapes formed from other materials?
19. What is meant by *green stage* as applied to work in fibreglass and resins?
20. Name four types of sandpaper, explaining the advantages and disadvantages of each type named.
21. What is *dip powder-coating*? Explain in detail how you would dip powder-coat a metal article which you had just completed.
22. How would you clean a brush which you had just finished using for applying several coats of paint? Why is it important to clean the brush when you have finished working with it?
23. Explain, with sketches, the following terms associated with the turning of metal:
(a) top slide; (b) tailstock; (c) faceplate; (d) bent-leg carrier.
24. What are the differences between a three-jaw self-centreing chuck and a four-jaw independent chuck? Give examples where each chuck should be used.
25. Explain, with sketches, the differences between the cutting actions of the following turning woodwork tools:
(a) turning gouge; (b) skew turning chisel; (c) scraper chisel.
26. Sketch the cutting ends of the three following lathe tools:
(a) roughing; (b) right-hand; (c) round-nosed.
27. What is meant by the terms rake angle and clearance angle in connection with lathe tools? State the rake and clearance angles for lathe tools when turning:
(a) cast iron; (b) mild steel; (c) nylon.
28. How would you harden and temper a centre punch?
21. How would you anneal (a) brass; (b) copper; (c) aluminium?
30. Why do plane cutters have separate grinding and cutting angles? What are these angles? How would you sharpen a plane cutter?
31. What differences are there between sharpening (a) a plane cutter; (b) a spokeshave cutter; (c) a firmer chisel?
32. Explain, with the aid of sketches, how you would sharpen a centre bit.
33. A soldering iron should be kept 'tinned'. Explain what this sentence means. How is a soldering iron tinned? What is the purpose of 'tinning'?

Index